KT-210-048

Stars

Ian Ridpath,
illustrations by Wil Tirion

Ian Ridpath and Wil Tirion are also authors of the *Collins Pocket Guide to Stars and Planets* (HarperCollins) and *The Monthly Sky Guide* (CUP). Ian Ridpath is editor of *Norton's Star Atlas* and the *Oxford Dictionary of Astronomy*, and author of *Star Tales*. Wil Tirion is author of *Sky Atlas 2000*.
www.ianridpath.tk
www.wil-tirion.tmfweb.nl

HarperCollins*Publishers*
Westerhill Road, Bishopbriggs, Glasgow G64 2QT

First published as *Gem Night Sky* 1985
This edition published as *Gem Stars* 2004

Reprint 10 9 8 7 6 5 4 3 2 1 0

ISBN 0-00-717858-1

Typesetting and layout by Ian Ridpath
Printed in Italy by Amadeus S.p.A.

Contents

How to use this book

This book contains individual charts of the 88 constellations that fill the entire sky, arranged alphabetically. All the stars marked on the charts are visible to the naked eye under the clearest skies, but in towns, where the skies are polluted by dirt and the glare from streetlights, binoculars will be necessary to pick out the faintest stars shown. In addition to the naked-eye stars, the maps mark the location of other important objects such as star clusters, nebulae and galaxies; most of these require some form of optical aid to be seen.

All the constellations are shown to the same scale, with six exceptions – Eridanus, Hercules, Hydra, Leo, Pisces and Ursa Major – which are drawn to slightly smaller scales. Specific areas of interest in certain constellations (such as the Pleiades star cluster, p. 229) are shown in greater detail on larger-scale maps which include fainter stars than the main charts.

Each constellation chart is accompanied by descriptions of the main objects of interest to the amateur observer, with an indication of the type of instrument and the magnification needed to see them. The majority of objects described are within the range of small telescopes (i.e. telescopes whose front lens has an aperture of 50–60 mm).

SEASONAL MAPS

This book is usable anywhere in the world, although not all constellations are visible from any given place, e.g. constellations in the south polar region of the sky will always remain below the horizon for mid-northern latitude observers, and vice versa. In addition, the constellations visible from any given latitude change with the seasons. The appearance of the sky at different times of the year is shown for a range of northern and southern latitudes on pp. 28–43.

ALL-SKY MAPS

The position of the constellations in relation to each other can be found from a ten-page mini-atlas of the entire sky on pp. 44–53.

Introduction

The sky is divided up into 88 areas, known as *constellations*, which serve as a convenient way of locating objects in the sky. The stars of a constellation usually have no physical connection with each other: although they appear in the same direction in the sky, they are actually at vastly differing distances from us.

Constellations come in many different shapes and sizes. The largest constellation, Hydra, the water snake, is a long and rambling figure that covers an area of sky 19 times greater than that of the smallest constellation,

Crux, the southern cross. Some constellations, such as Orion, consist of easily recognizable patterns of bright stars, while others are faint and difficult to identify.

The tradition of dividing the sky into constellations began thousands of years ago when ancient peoples assigned certain star patterns the names of their gods, heroes and fabled animals. With few exceptions, the star patterns bear very little resemblance to the people and creatures they are supposed to represent; the connections are symbolic rather than literal.

The ancient Greeks recognized a total of 48 constellations, including the 12 constellations of the *zodiac*, through which the Sun passes on its yearly path around the sky. (The Sun's path, known as the *ecliptic*, is plotted on the maps in this book as a red dashed line.) Various other constellations were added at later times.

Early celestial cartographers drew the constellation figures as they pleased, for there was no standardized shape for each one, nor was there even a generally agreed list of constellations. Each cartographer was free to introduce new constellations of his own invention, and to amend or omit the inventions of others. Constellation figures frequently overlapped one another, and sometimes stars were shared between two constellations.

This state of confusion persisted until 1930 when the International Astronomical Union, astronomy's governing body, adopted the list of 88 constellations that we know today, and set down their exact boundaries. There is no particular reason why we should have 88 constellations, or why they should be the shape they are. Rather like the political map of the world, the sub-division of the sky is an accident of history. But, unlike the countries of the world, the names and borders of the constellations are not likely to change until the slow, steady movements of the stars, known as their *proper motions*, render the existing constellation shapes unrecognizable, thousands of years from now.

STAR BRIGHTNESSES

Star brightnesses are expressed in terms of *magnitudes* (abbreviated mag.). This system was started by the Greek astronomer Hipparchus in the second century BC. He divided the stars into six categories of brightness, from the brightest stars (first magnitude) to the faintest that he could see (sixth magnitude).

Nowadays star brightnesses are measured to the nearest hundredth of a magnitude by sensitive instruments known as photometers. A difference of five magnitudes is defined as being exactly equal to a brightness difference of 100 times; consequently, each magnitude step is equal to the fifth root of 100, which is approximately 2.5. Stars more than 2.5 times brighter than mag. 1 are given negative (minus) magnitudes, e.g. Sirius, the brightest star in the sky, which has a magnitude of −1.44. Stars fainter than 6th mag. are given progressively larger positive magnitudes. The faintest objects seen through telescopes on Earth have magnitudes of about 25. The table on the facing page shows the difference in brightness that corresponds to a given magnitude difference.

The charts of individual constellations in this book include all stars down to mag. 6.0; the total number of stars depicted is approximately 5000. The all-sky atlas shows stars down to mag. 5.0, while the seasonal charts show only the brighter stars, to mag. 4.5.

The magnitude of diffuse objects such as star clusters, nebulae, galaxies, and comets is more difficult to quantify than that of a star. A diffuse object's brightness is usually assessed as though all its light were concentrated into one star-like point. Therefore a galaxy of 9th mag. appears the same brightness as would an out-of-focus star of 9th mag. whose light was spread over the same area as the galaxy. The best way to estimate the brightness of diffuse objects is to compare them with the image of an out-of-focus star of known magnitude.

STAR NAMES

There are several different systems for identifying stars, and as a result a given star may be referred to in more than one way. Many of the brightest stars have proper names which can be of Arabic, Greek or Latin origin; examples are Altair, Sirius, and Regulus respectively. Another common system is to give the brightest stars in each constellation a Greek letter; this system was started in 1603 by the German celestial cartographer

MAGNITUDE DIFFERENCE CONVERTED TO BRIGHTNESS DIFFERENCE

Difference in magnitude	Difference in brightness
0.5	1.6
1.0	2.5
1.5	4.0
2.0	6.3
2.5	10
3.0	16
3.5	25
4.0	40
5.0	100
6.0	250
7.5	1000
10	10,000
12.5	100,000
15	1,000,000

Johann Bayer, so that the Greek letters attached to stars are known as *Bayer letters*. For example, Sirius is also known as α (alpha) Canis Majoris, meaning that it is the star α (alpha) in the constellation Canis Major (note that the genitive form of the constellation's name is always used when referring to a star in such a context).

Stars that are not identified in either of these ways may be known by their *Flamsteed number* (e.g. 61 Cygni) given to stars that were listed in a catalogue compiled by the first English Astronomer Royal, John Flamsteed (1646–1719). Fainter stars are referred to by their numbers in any one of several other star catalogues.

Various astronomers have also compiled specialized lists of particular types of stars – such as double stars, nearby stars, or white dwarf stars – and these provide another source of nomenclature. In this book, some double stars catalogued by the Russian astronomer F. G. W. Struve are referred to by their Struve numbers (these numbers are sometimes preceded by the symbol Σ, the capital Greek letter S, for Struve).

Variable stars have a nomenclature all their own. Those that are not named under the existing systems are given one or two Roman letters (e.g. W Virginis, RR Lyrae). When all possible letter combinations have been used up, the variable stars in a constellation are denoted by the letter V and a number, e.g. V 1500 Cygni.

THE GREEK ALPHABET

α	alpha	ι	iota	ρ	rho
β	beta	κ	kappa	σ	sigma
γ	gamma	λ	lambda	τ	tau
δ	delta	μ	mu	υ	upsilon
ε	epsilon	ν	nu	φ	phi
ζ	zeta	ξ	xi	χ	chi
η	eta	ο	omicron	ψ	psi
θ	theta	π	pi	ω	omega

Objects such as star clusters, nebulae and galaxies have their own systems of nomenclature, the most familiar of which are the M and NGC numbers. The M numbers come from a catalogue of over 100 clusters and nebulous objects compiled in the 18th century by the French astronomer Charles Messier. The NGC numbers come from the *New General Catalogue of Nebulae and Clusters of Stars* published in 1888 by the Danish astronomer J. L. E. Dreyer. Two supplements to the NGC, called the *Index Catalogues*, appeared in 1895 and 1908; objects listed in these are given IC numbers. Most objects with M numbers also have NGC numbers.

STAR TYPES

Stars are incandescent balls of gas, similar in nature to the Sun but so far away that they appear as nothing more than points of light in even the largest telescopes. Nevertheless, by analysing the light from the stars, astronomers have been able to deduce that stars come in a wide range of sizes, temperatures, colours and brightnesses.

The largest and brightest stars, aptly termed *giants* and *supergiants*, are hundreds of times the diameter of the Sun, so that they would encompass the orbit of the Earth if they were placed where our Sun is. Stars of such enormous girth are at a more advanced stage of evolution than our own Sun, which itself will swell up into a red giant towards the end of its life.

At the other end of the scale are the *red dwarf* stars that have about one-tenth the diameter of the Sun; these are stars that were born with much less mass than the Sun. Most remarkable of all are the super-dense *white dwarf* stars which have the mass of the Sun packed into a sphere the size of the Earth. These are thought to be the exposed cores of former red giant stars whose outer layers of rarefied gas have dissipated into space.

STAR COLOURS

Not all stars are white as they at first appear. More careful inspection reveals that stars come in a wide range of colours, from deep orange through yellow to blue-white. The colour of a star depends on its surface temperature, with the coolest stars being the reddest and the hottest ones the bluest. Therefore a star's colour is a clue to its physical nature.

The stars with the most prominent colours are the red giants and supergiants such as Betelgeuse in Orion, Aldebaran in Taurus and Antares in Scorpius. Star colours are more prominent through binoculars and telescopes than with the naked eye. Particularly attractive are double stars (p. 17) in which the stars are of contrasting colours.

On the constellation maps in this book, the first-magnitude stars are coloured to give an indication of their surface temperatures, rather than the exact colour they appear to the eye. The fainter the star, the less likelihood there is of seeing any colour in it.

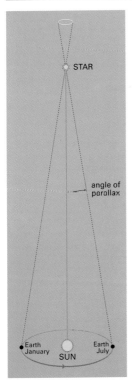

STAR DISTANCES

Distances to stars and galaxies are usually measured either in *light years* (abbreviated l.y.) or *parsecs*; in this book the light year is used. It is the distance that a beam of light, moving at 300,000 km per second, travels in one year. One l.y. is equivalent to 9.5 million million km.

The distances of the nearest stars can be measured directly by a technique known as *parallax*. This involves measuring the star's precise position against the celestial background, as seen from opposite sides of the Earth's orbit. A shift in the star's position when seen from the two viewpoints (on either side of the Sun, six

Parallax, a star's change in position as seen from opposite sides of the Earth's orbit, reveals the star's distance

months apart) reveals the star's distance, with the nearest stars to us showing the greatest amount of parallax shift. A star at a distance of 3.26 l.y. would show a parallax shift of one second of arc; hence a distance of 3.26 l.y. is known as a parsec (short for parallax of one second). In practice, no star is quite this close. The nearest star to us, Proxima Centauri, has a parallax of 0.77 seconds of arc, which makes its distance 4.2 l.y.

Because parallaxes are so small, only the distances of the closest stars could be measured in this way by telescopes on the ground. Now, however, a satellite called Hipparcos has accurately measured the parallaxes of over 100,000 stars from space, and those distances are used in this book.

For stars which are too distant to show a measurable parallax even to Hipparcos, astronomers use an indirect method that involves estimating the star's luminosity from features of its spectrum; they can then work out how far away from us the star must be to account for its brightness as seen in the sky. This method is open to considerable error, but before Hipparcos it was the only technique available for finding the distance of the majority of stars.

VARIABLE STARS

Some stars are not constant in brightness, but vary over periods of time ranging from hours to weeks or even years. The brightness of a variable star can be estimated by comparison with surrounding stars that do not vary.

The most common cause of variation is that the star actually pulsates in size as a result of an instability. A celebrated class of pulsating variables is the Cepheids, named after their prototype, δ (delta) Cephei (p. 96). These are yellow supergiant stars that pulsate regularly every few days or weeks, depending on their brightness.

Their importance to astronomers is that their period of pulsation is directly related to their luminosity, the brightest Cepheids taking the longest to pulsate. Consequently, by observing a Cepheid's pulsation period astronomers can accurately ascertain its luminosity. And by comparing the calculated luminosity with the star's brightness as it appears on Earth, they can then work out how far away it is. Cepheids are thus an important tool for calibrating distances in the Universe.

For all their importance, Cepheids are relatively rare. The most abundant type of variable stars are actually the red giants and supergiants, virtually all of which exhibit some form of variability due to pulsations in size, although they do not have the strict periodicity of Cepheids; a famous example of a red giant variable is o (omicron) Ceti, popularly known as Mira (see p. 100). Some red variable stars, such as the supergiant Betelgeuse, follow no detectable pattern at all.

A totally different type of variable star is the eclipsing binary. It consists of two stars in mutual orbit, one periodically moving in front of the other as seen from Earth. Each eclipse of one star by the other causes a dip in the total light that we receive. The most famous eclipsing binary star is Algol, also known as β (beta) Persei (see p. 190).

Most spectacular of all are the eruptive variables which undergo sudden and often very large changes in light output, most notably the novae and supernovae. A *nova* is thought to be a close double star in which one member is a white dwarf. Gas from the companion star spills onto the white dwarf where it ignites explosively, causing the star's light to surge temporarily by thousands of times. The star is not destroyed in a nova explosion; some novae have been seen to erupt more than once, and possibly all novae may recur given time. Novae are often first spotted by amateur astronomers.

Even more spectacular than normal novae are the *supernovae*, celestial cataclysms that signal the death of a star. In a supernova, the star ends its life by blowing itself to bits, temporarily shining as brilliantly as billions of normal stars. Where the supernova occurred, the wreckage of the star is left to drift away into space, as with the Crab Nebula in Taurus (p. 230) and the Veil Nebula in Cygnus (p. 118).

Supernovae come in two distinct types. One sort involves the explosion of a white dwarf in a binary system. In the other sort, a star many times more massive than the Sun becomes unstable and explodes. The last supernova in our Galaxy was seen in 1604 and we are long overdue for another, although a supernova flared up to naked-eye brightness in the Large Magellanic Cloud in 1987.

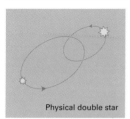

Physical double star

Optical double star

DOUBLE AND MULTIPLE STARS

When examined through telescopes, many stars are found not to be single, as they appear to the naked eye, but to be accompanied by one or more companions. In some cases the stars are not really connected but simply happen to lie in the same line of sight by chance; such an arrangement is known as an *optical double*.

But in the great majority of cases the stars are linked by gravity and orbit around each other, the exact time taken depending on their distance apart. A pair of stars linked in this way is known as a *binary*; sometimes whole families of

In a physical double, two stars orbit each other; in an optical double, the stars lie in the same line of sight but at different distances

three or more stars may be interlinked by gravity, giving rise to a *multiple star*.

The apparent separation of two stars in a double is measured in seconds of arc. There are 60 seconds (″) in a minute (′) of arc, and 60 minutes of arc in a degree (°), so that one second of arc is 1/3600th of a degree. The apparent diameter of the Moon is about 30 minutes of arc, or half a degree.

The widest doubles – those with separations of several seconds of arc or more – can be divided in small telescopes or even binoculars, but the closer together the components of a double star are, the larger the aperture of telescope needed to separate them. The steadiness of the air, known as the *seeing*, will also affect the ability to separate close doubles. The best examples of observable doubles are described in the text accompanying each map.

A *spectroscopic binary* is a pair that is too close together to be separated in a telescope, although the existence of a second star is revealed when the light is analysed in a device called a spectroscope.

STAR CLUSTERS

Stars sometimes group together in clusters, of which there are two main types: open clusters and globular clusters. *Open clusters* are the less densely packed of the two types: they are usually irregular in shape, and contain anything up to many thousands of relatively young stars. They lie in the spiral arms of our Galaxy. Famous examples are the Pleiades and Hyades clusters in Taurus (pp. 228–229), and the Double Cluster in Perseus (p. 192). Open clusters frequently cover half a degree of sky or more, i.e. equal to or greater than the apparent diameter of the full Moon.

Globular clusters are dense, ball-shaped aggregations that can contain hundreds of thousands of stars. They are distributed in a halo around our Galaxy, so they are usually much farther away from us than open clusters and in general they appear smaller and are more difficult to resolve into individual stars. In contrast to open clusters, globular clusters contain some of the oldest stars known. Famous examples are ω (omega) Centauri (p. 94), 47 Tucanae (p. 234), and M13 in Hercules (p. 138).

NEBULAE

Nebulae are clouds of gas and dust in space, some of which glow brightly while others are dark. In bright nebulae, the gas and dust is lit up by the stars that lie within them. Famous examples of bright nebulae are the Orion Nebula (p. 184) and the Tarantula Nebula (in Dorado, p. 122). By contrast, dark nebulae are visible only because they blot out light from objects behind them. One famous dark nebula is the Coalsack in Crux, the Southern Cross, which obscures part of the Milky Way star fields of that region. Some dark nebulae are seen in silhouette against bright nebulae.

PLANETARY NEBULAE

Planetary nebulae are clouds of gas that have been
thrown off by a central star at the end of its life. Their
name is misleading, for planetary nebulae have nothing
to do with real planets at all. Rather, the name was
given because their appearance often resembles the disc
of a planet as seen through small telescopes.

MILKY WAY

All the stars visible to the naked eye are part of an
enormous system of at least 100,000 million stars
known as the Galaxy (our Galaxy is given a capital

*The Lagoon Nebula, M8, in Sagittarius, contains the open
cluster NGC 6530 (on the left in this photograph). The
red colour of the nebula, prominent on photographs, is not
apparent to the naked eye. AURA/NOAO/NSF*

letter to distinguish it from other galaxies – see below). Those stars nearest to us in the Galaxy are scattered more or less at random across the sky and make up the constellations. The more distant ones mass into a faint band of light seen crossing the sky on clear nights. This is the Milky Way, and is shown on the constellation maps in this book as a light blue band. Sometimes the term Milky Way is used as an alternative name for our Galaxy as a whole. Our Galaxy has a spiral shape, with most of the stars and nebulae concentrated in curving arms that wind outwards from a bulging central core. The Galaxy is about 100,000 l.y. in diameter and the Sun lies 30,000 l.y. from its centre. The centre of our Galaxy lies in the direction of the constellation Sagittarius, a region in which the Milky Way star fields are particularly dense. The entire Galaxy is rotating; our Sun takes about 250 million years to complete one orbit around the centre of the Galaxy.

GALAXIES

Galaxies extend into space as far as the largest telescopes can see. Each is a collection of millions or billions of stars held together by the mutual attraction of gravity. Galaxies are classified according to their shapes, of which there are two main forms: spiral and elliptical.

Spiral galaxies are sub-divided into normal spirals and barred spirals. Most galaxies within reach of amateur telescopes are normal spirals, with a central bulge of stars from which arms curve outwards. The most famous example is the Andromeda Galaxy (see pp. 56–57). *Barred spirals* differ by having a straight bar of stars across their centres; the spiral arms emerge from the ends of this bar. Our own Galaxy is usually classified as a normal spiral, although there is some evidence that it might be a barred spiral. Spiral galaxies are oriented at different angles with respect to us, which affects the way they appear in the telescope. For instance, a spiral seen face-on appears circular in telescopes, but an edge-on spiral appears cigar-shaped.

Elliptical galaxies come in a wide range of sizes, from the most massive to the smallest galaxies of all. They are elliptical in outline and have no spiral arms. There are also a few galaxies that are irregular in shape. These include the two Magellanic Clouds, which are satellite galaxies of our Milky Way, although there is a trace of spiral form in the Large Magellanic Cloud.

Since galaxies are faint and fuzzy, they are best viewed with low magnification to increase the contrast against the sky background.

STAR POSITIONS AND MOTIONS

In the sky, the equivalents of longitude and latitude are right ascension (R.A.) and declination (dec.). *Right ascension* starts at the place where the Sun moves northwards across the celestial equator each year. This point, known as the *vernal equinox*, is the celestial equivalent of the Greenwich meridian on Earth.

Right ascension is measured eastwards from the vernal equinox around the sky in hours, from 0h to 24h. Each hour of right ascension is divided into 60 minutes, and each minute is divided into 60 seconds. *Declination* is measured in degrees north and south of the celestial equator, from 0° at the equator to +90° at the north celestial pole and −90° at the south celestial pole. The celestial poles lie directly above the Earth's poles, and the celestial equator is directly overhead as seen from the Earth's equator.

The position of a star or any other object can therefore be specified precisely in terms of right ascension and declination, like the coordinates of a place on Earth. Coordinate grids marked in hours of right ascension and degrees of declination are overlain on the star maps in this book.

But celestial cartographers are faced with two problems that do not concern their terrestrial counterparts. Firstly, each star is moving very slowly relative to its neighbours; this is known as *proper motion*. With few exceptions, e.g. Barnard's Star (see pp. 178–179), the proper motions of stars are so small as to be undetectable over a human lifetime except by precise measurements. However, over many thousands of years these motions will build up until the present shapes of the constellations are changed totally, and stars will have strayed into neighbouring constellations. One day, astronomers will have to revise the current nomenclature of stars and constellations.

The second problem is that the entire coordinate grid is itself shifting due to a wobble of the Earth in space called *precession*. This causes the zero point of right ascension to move once around the sky every 26,000 years. As a result, the coordinates of all points in the sky are gradually changing, and it is customary to give the coordinates of celestial objects with respect to a certain reference date. In the case of this book the reference date is the year 2000, so the star maps herein will be usable without serious error until the middle of the 21st century.

PLANETS

In addition to the fixed stars, the planets of our Solar System are also visible in the night sky. The planets shine by reflecting light from the Sun. Because they are always moving as they travel along their orbits around the Sun, the planets cannot be shown on the maps in this book. But when the planets do appear, they will be found near the ecliptic.

The brightest planet, Venus, outshines every star in the sky. It can be seen either rising before the Sun in the morning sky, when it is popularly known as the morning star, or setting in the evening twilight, when it is termed the evening star. A small telescope will reveal that Venus is not a star at all, for it shows a disk that goes through phases, like those of the Moon, as it orbits the Sun. The second-brightest planet is Jupiter, which at its brightest also easily outshines any star.

Unlike Venus, which keeps close to the Sun, Jupiter can be seen in any part of the sky. Binoculars, if held steadily, reveal its four brightest moons that circle it incessantly (Io, Europa, Ganymede, Callisto), changing their positions from night to night.

Mars and Saturn can also feature prominently in the night sky. Mars is notable because of its glaring red colour, but it is a disappointing sight in amateur telescopes because it is so small. Saturn, by contrast, is perhaps the most beautiful telescopic sight of all, being encircled by flat rings composed of snowballs that orbit it like a swarm of tiny moons. Of the other planets, Mercury is an elusive object that never strays far enough from the Sun to be easily visible, while Uranus, Neptune, and Pluto are distant and faint.

METEORS

Occasionally a bright streak of light will be seen darting across the sky, lasting no more than a second or so. This is a *meteor*, popularly known as a *shooting star*. Meteors have nothing to do with stars. They are actually specks of dust burning up by friction as they dash at high speed into the Earth's upper atmosphere. We do not see the speck of dust itself, but rather the trail of hot gas that it produces as it burns up at a height of about 100 km.

On any clear night a few meteors can be seen with the naked eye each hour; these random arrivals are known as *sporadic* meteors. But several times each year the

Earth passes through a swarm of interplanetary dust, which gives rise to a *meteor shower*. The members of a meteor shower all appear to come from one small area of sky, known as the *radiant*. The shower is named after the constellation in which the radiant lies, e.g. the Geminids appear to radiate from Gemini. One exception is the Quadrantids, which radiate from a part of the sky once occupied by the now-defunct constellation Quadrans Muralis, the mural quadrant, now part of Boötes. A meteor shower may last for days or even weeks as the Earth passes through the swarm of dust, but the peak of activity is usually confined to one particular night.

In the case of the richest showers, such as the Perseids in August, dozens of meteors per hour may be seen at best. Sometimes the meteors explode along their path, becoming bright enough to cast shadows. Some meteors leave trains that slowly fade.

The number of meteors in a shower is expressed in terms of the *zenithal hourly rate* (ZHR). This is the number of meteors that an observer would see each hour if the radiant were directly overhead. In practice it seldom is, and so the actual number of meteors seen will be less than the theoretical ZHR. Also, if conditions are less than perfect – e.g. polluted skies or the presence of bright moonlight – then the number of meteors seen will again be reduced. The main meteor showers of the year are listed in the table on the facing page, with their dates of peak activity and ZHR.

MAJOR ANNUAL METEOR SHOWERS

Shower	Date(s) of maximum	ZHR
Quadrantids	January 3–4	100
Lyrids	April 21–22	10
Eta Aquarids	May 5	35
Delta Aquarids	July 28–29	20
Perseids	August 12–13	80
Orionids	October 20–22	25
Taurids	November 4	10
Leonids	November 17–18	10
Geminids	December 13–14	100

Key to the symbols used on the constellation maps

Magnitudes

brighter than –0.5	1.1 – 1.5	3.6 – 4.0
–0.5 – 0.0	1.6 – 2.0	4.1 – 4.5
0.1 – 0.5	2.1 – 2.5	4.6 – 5.0
0.6 – 1.0	2.6 – 3.0	5.1 – 5.5
	3.1 – 3.5	5.6 – 6.0

Double stars   Variable stars

Open clusters   Globular clusters

Diffuse nebulae   Planetary nebulae

Galaxies

1 – Winter sky

WEST

ECLIPTIC

PISCES

PEGASUS

ARIES

TRIANGULUM

LACERTA

ANDROMEDA

Pleiades

CYGNUS

Deneb

CEPHEUS

PERSEUS

Algol

Zenith 40°N

Capella

LYRA

CASSIOPEIA

Zenith 60°N

AURIGA

Vega

Horizon 60°N

GEMINI

Castor

Horizon 40°N

CAMELOPARDALIS

Pollux

NORTH

HERCULES

DRACO

Horizon 20°N

Polaris

URSA MINOR

LYNX

CANCER

LEO MINOR

DRACO

CORONA BOREALIS

URSA MAJOR

CANES VENATICI

LEO

Regulus

BOOTES

COMA BERENICES

ECLIPTIC

VIRGO

Arcturus

EAST

Magnitudes

0   1   2

Northern latitudes

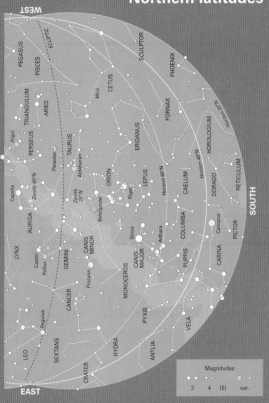

2 – Spring sky

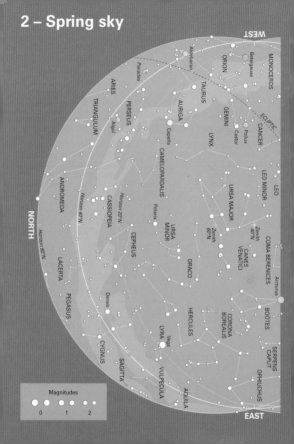

Northern latitudes

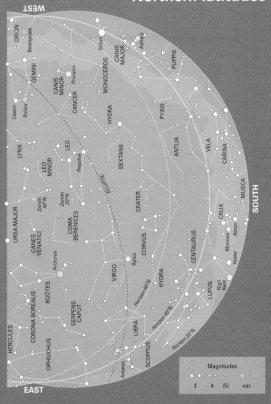

3 – Summer sky

VIRGO

PEGASUS

LEO
COMA BERENICES

LEO MINOR

CANCER

CANES VENATICI

BOÖTES

Arcturus

Pollux
Castor
GEMINI

URSA MAJOR

CORONA BOREALIS

LYNX

Horizon 60°N

Horizon 40°N

Horizon 20°N

DRACO

URSA MINOR

Zenith 60°N

Zenith 40°N

HERCULES

LYRA

Vega

AURIGA

Capella

CAMELOPARDALIS

Polaris

CYGNUS

Deneb

VULPECULA

NORTH

PERSEUS

CASSIOPEIA

CEPHEUS

LACERTA

PEGASUS

Algol

TRIANGULUM

ANDROMEDA

ARIES

PISCES

PISCES

Magnitudes

0   1   2

Northern latitudes

WEST

LEO
COMA BERENICES
CORVUS
HYDRA
Spica
VIRGO
BOOTES
Arcturus
CENTAURUS
Horizon 20°N
Horizon 40°N
Horizon 60°N
DRACO
CORONA BOREALIS
SERPENS CAPUT
LIBRA
Rigil Kent
Zenith 40°N
Zenith 20°N
HERCULES
OPHIUCHUS
Antares
LUPUS
CIRCINUS
Vega
LYRA
SERPENS CAUDA
SCORPIUS
ARA
TRIANGULUM AUSTRALE
CYGNUS
SCUTUM
CORONA AUSTRALIS
Deneb
VULPECULA
SAGITTA
AQUILA
SAGITTARIUS
TELESCOPIUM
PAVO
Altair
EQUULEUS DELPHINUS
CAPRICORNUS
PEGASUS
ECLIPTIC
MICROSCOPIUM
INDUS
GRUS
AQUARIUS
PISCES
PISCIS AUSTRINUS
Fomalhaut

SOUTH

EAST

Magnitudes

3   4   (5)   var.

4 – Autumn sky

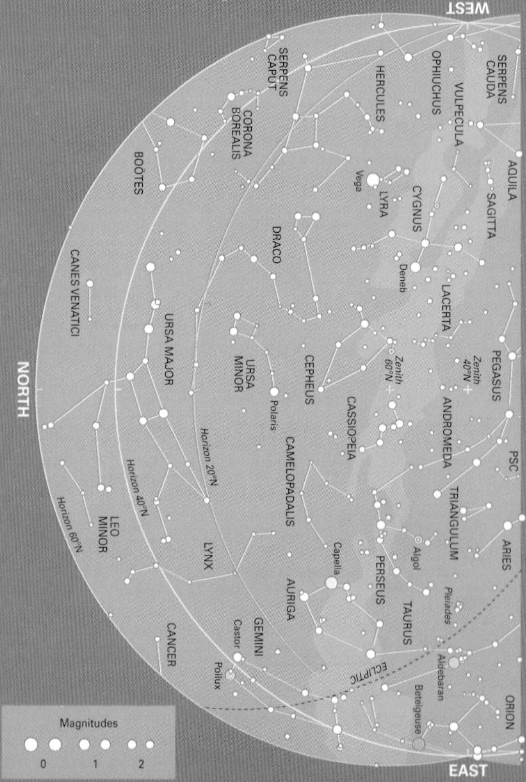

WEST

NORTH

EAST

SERPENS CAUDA
SERPENS CAPUT
VULPECULA
OPHIUCHUS
HERCULES
AQUILA
SAGITTA
CORONA BOREALIS
BOÖTES
CYGNUS
Vega
LYRA
DRACO
Deneb
LACERTA
PEGASUS
Zenith 40°N
CANES VENATICI
URSA MAJOR
URSA MINOR
Polaris
CEPHEUS
Zenith 60°N
ANDROMEDA
PSC
TRIANGULUM
ARIES
CASSIOPEIA
Algol
Perseus
Pleiades
CAMELOPARDALIS
Horizon 20°N
Horizon 40°N
LYNX
Capella
AURIGA
Aldebaran
TAURUS
Horizon 60°N
LEO MINOR
CANCER
Castor
Pollux
GEMINI
Betelgeuse
ECLIPTIC
ORION
PERSEUS

Magnitudes

0   1   2

Northern latitudes

WEST

EAST

SOUTH

OPHIUCHUS
SERPENS CAUDA
SCUTUM
VULPECULA
SAGITTARIUS
CORONA AUSTRALIS
AQUILA
CYGNUS
SAGITTA
Altair
DELPHINUS
CAPRICORNUS
MICROSCOPIUM
INDUS
Deneb
LACERTA
EQUULEUS
PEGASUS
Zenith 40°N
Zenith 20°N
GRUS
PAVO
ANDROMEDA
AQUARIUS
PISCIS AUSTRINUS
Fomalhaut
PHOENIX
TUCANA
Algol
PERSEUS
TRIANGULUM
PISCES
ECLIPTIC
SCULPTOR
Horizon 20°N
HYDRUS
ARIES
CETUS
Horizon 60°N
Horizon 40°N
Achernar
Aldebaran
Pleiades
Mira
FORNAX
HOROLOGIUM
TAURUS
ERIDANUS
Betelgeuse
ORION
LEPUS
Rigel

Magnitudes

3 4 (5) var.

5 – Summer sky

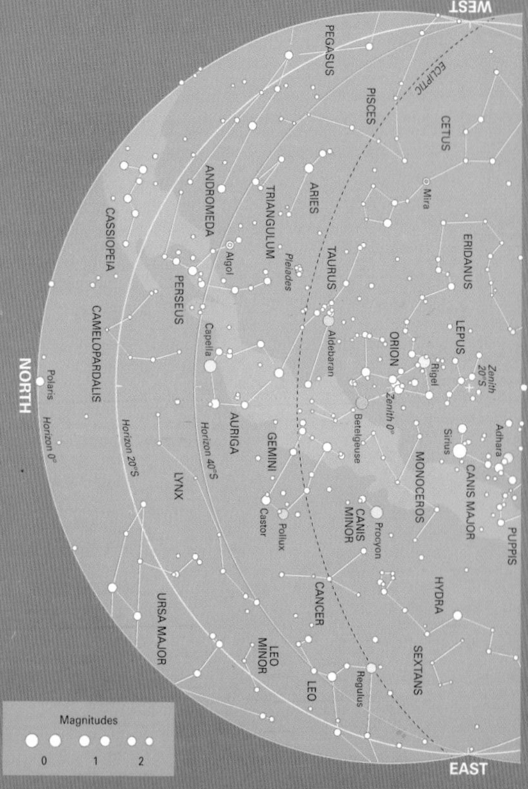

WEST

PEGASUS
PISCES
ECLIPTIC
CETUS
Mira
ANDROMEDA
ARIES
TRIANGULUM
ERIDANUS
Algol
Pleiades
TAURUS
CASSIOPEIA
PERSEUS
Aldebaran
LEPUS
ORION
Zenith 20°S
Rigel
Zenith 0°
Betelgeuse
Sirius
Antares
CAMELOPARDALIS
Capella
AURIGA
Betelgeuse
CANIS MAJOR
NORTH
Polaris
Horizon 0°
Horizon 20°S
Horizon 40°S
GEMINI
MONOCEROS
PUPPIS
LYNX
Castor
Pollux
CANIS
MINOR
Procyon
HYDRA
URSA MAJOR
CANCER
SEXTANS
LEO
MINOR
LEO
Regulus

Magnitudes
0   1   2

EAST

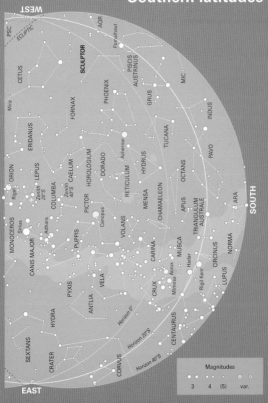

Southern latitudes

WEST

PSC

ECLIPTIC

AQR

Fomalhaut

CETUS

SCULPTOR

PISCIS AUSTRINUS

Mira

PHOENIX

MIC

FORNAX

GRUS

ERIDANUS

INDUS

Achernar

ORION

Rigel

LEPUS

Zenith 20°S

CAELUM

HOROLOGIUM

RETICULUM

HYDRUS

TUCANA

COLUMBA

Zenith 40°S

DORADO

PAVO

MONOCEROS

PICTOR

VOLANS

MENSA

CHAMAELEON

OCTANS

Sirius

Canopus

ARA

Adhara

CANIS MAJOR

PUPPIS

CARINA

MUSCA

APUS

TRIANGULUM AUSTRALE

SOUTH

PYXIS

VELA

CRUX

Mimosa Acrux

Hadar

CIRCINUS

NORMA

ANTLIA

Rigil Kent

LUPUS

HYDRA

Horizon 0°

CENTAURUS

SEXTANS

Horizon 20°S

CRATER

CORVUS

Horizon 40°S

EAST

Magnitudes

3 4 (5) var.

6 – Autumn sky

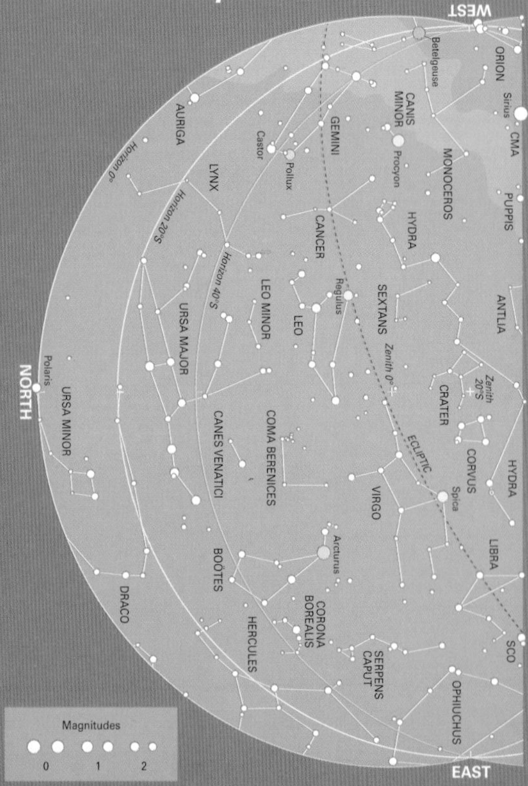

WEST

NORTH

EAST

ORION
Betelgeuse
CMA
Sirius
CANIS MINOR
Procyon
MONOCEROS
PUPPIS
GEMINI
Castor
Pollux
HYDRA
ANTLIA
Horizon 60°S
AURIGA
LYNX
CANCER
Horizon 20°S
SEXTANS
Zenith 0°
Zenith 20°S
CRATER
LEO MINOR
Regulus
LEO
CORVUS
ECLIPTIC
HYDRA
Horizon 40°S
URSA MAJOR
CANES VENATICI
COMA BERENICES
Spica
VIRGO
LIBRA
Polaris
URSA MINOR
DRACO
BOOTES
Arcturus
CORONA BOREALIS
SCO
HERCULES
SERPENS CAPUT
OPHIUCHUS

Magnitudes
0   1   2

Southern latitudes

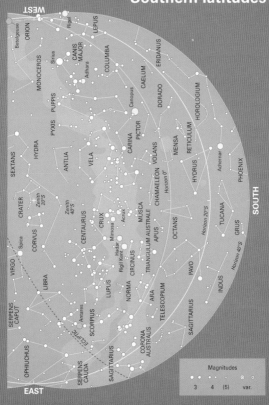

WEST

ORION
Betelgeuse
Rigel
LEPUS
MONOCEROS
CANIS MAJOR
Sirius
Adhara
COLUMBA
ERIDANUS
CAELUM
Canopus
DORADO
HOROLOGIUM
PUPPIS
PYXIS
HYDRA
SEXTANS
ANTLIA
VELA
CARINA
PICTOR
RETICULUM
Horizon 0°
MENSA
HYDRUS
Achernar
PHOENIX
CRATER
Zenith 20°S
Zenith 40°S
CHAMAELEON
VOLANS
CORVUS
CENTAURUS
CRUX
Mimosa
Acrux
MUSCA
APUS
OCTANS
Horizon 20°S
TUCANA
GRUS
Spica
VIRGO
Hadar
Rigil Kent
CIRCINUS
TRIANGULUM AUSTRALE
Horizon 40°S
PAVO
INDUS
LIBRA
LUPUS
NORMA
ARA
TELESCOPIUM
Antares
SCORPIUS
SERPENS CAPUT
SAGITTARIUS
OPHIUCHUS
CORONA AUSTRALIS
SERPENS CAUDA
SAGITTARIUS
ECLIPTIC

SOUTH

EAST

Magnitudes

3   4   (5)   var.

7 – Winter sky

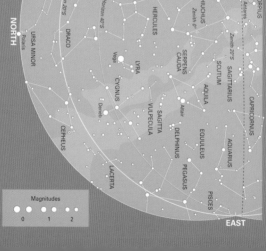

Southern latitudes

WEST

Horizon 40°S
S. 20° Horizon
Horizon 0°

Spica
VIRGO
ECLIPTIC
CORVUS
CRATER
ANTLIA
LIBRA
HYDRA
VELA
OPHIUCHUS
LUPUS
CENTAURUS
CRUX
Hadar
Mimosa
Acrux
MUSCA
CARINA
PUPPIS
SCORPIUS
NORMA
Rigil
Kent
TRIANGULUM
AUSTRALE
APUS
Antares
CIRCINUS
SOUTH
Canopus
PICTOR
SCUTUM
SER
Zenith 20°S
CORONA
AUST.
ARA
Zenith
40°S
PAVO
CHAMAELEON
MENSA
VOLANS
SAGITTARIUS
TELESCOPIUM
OCTANS
HYDRUS
DORADO
RETICULUM
CAPRICORNUS
MICROSCOPIUM
INDUS
TUCANA
HOROLOGIUM
AQUARIUS
GRUS
Achernar
ERIDANUS
PISCIS
AUSTRINUS
Fomalhaut
PHOENIX
SCULPTOR
PISCES
CETUS

EAST

Magnitudes

3   4   (5)   var.

8 – Spring sky

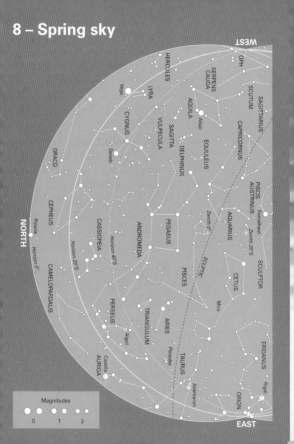

Southern latitudes

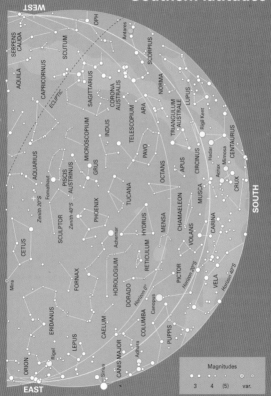

WEST

OPH

SERPENS
CAUDA

SCUTUM

Antares

AQUILA

CAPRICORNUS

ECLIPTIC

SCORPIUS

SAGITTARIUS

CORONA
AUSTRALIS

NORMA

ARA

LUPUS

TRIANGULUM
AUSTRALE

MICROSCOPIUM

INDUS

TELESCOPIUM

PAVO

Rigil Kent

AQUARIUS

GRUS

APUS

CIRCINUS

Acrux

Mimosa

Fomalhaut

PISCIS
AUSTRINUS

Zenith 20°S

TUCANA

OCTANS

MUSCA

CENTAURUS

CRUX

Zenith 40°S

PHOENIX

CHAMAELEON

CETUS

SCULPTOR

Achernar

HYDRUS

MENSA

VOLANS

CARINA

SOUTH

RETICULUM

FORNAX

HOROLOGIUM

DORADO

PICTOR

Horizon 20°S

VELA

Mira

Horizon 0°

CAELUM

Horizon 40°S

ERIDANUS

Canopus

LEPUS

PUPPIS

Rigel

COLUMBA

ORION

CANIS MAJOR

Adhara

ORION

Sirius

EAST

Magnitudes				
3	4	(5)		var.

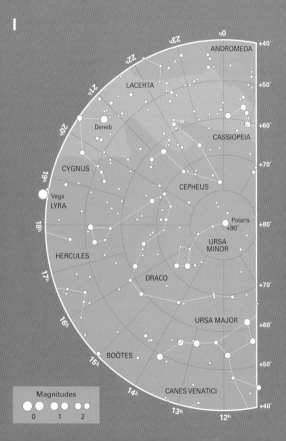

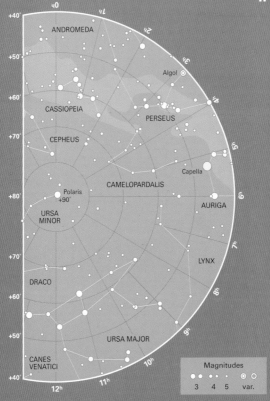

ANDROMEDA

+40°

+50°

+60°

+70°

+80°

+70°

+60°

+50°

+40°

1ʰ

2ʰ

3ʰ

Algol ⊙

4ʰ

CASSIOPEIA

PERSEUS

CEPHEUS

5ʰ

Capella

CAMELOPARDALIS

AURIGA

6ʰ

Polaris
+90°

URSA
MINOR

LYNX

7ʰ

DRACO

8ʰ

URSA MAJOR

9ʰ

CANES
VENATICI

10ʰ

11ʰ

12ʰ

Magnitudes

● ● ● · ⊙ ○
3   4   5   var.

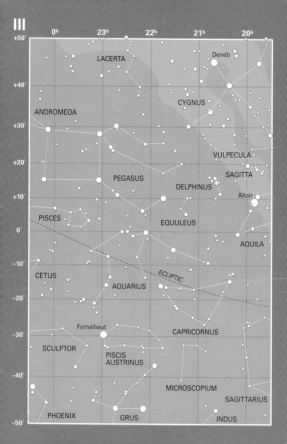

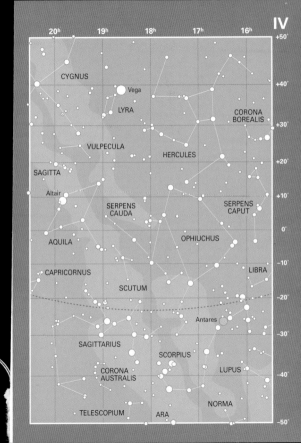

IV

20h 19h 18h 17h 16h +50°

CYGNUS

Vega

LYRA

CORONA
BOREALIS

+40°

+30°

VULPECULA

HERCULES

+20°

SAGITTA

Altair +10°

SERPENS
CAUDA

SERPENS
CAPUT

0°

AQUILA

OPHIUCHUS

-10°

CAPRICORNUS

LIBRA

SCUTUM

-20°

Antares

SAGITTARIUS

-30°

SCORPIUS

CORONA
AUSTRALIS

LUPUS

-40°

TELESCOPIUM

ARA

NORMA

-50°

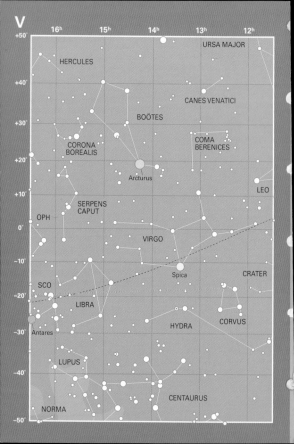

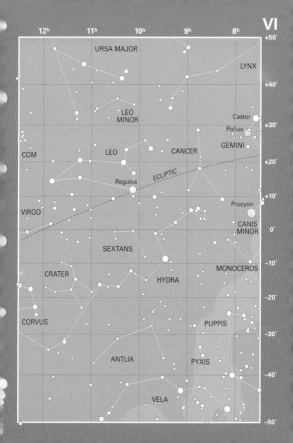

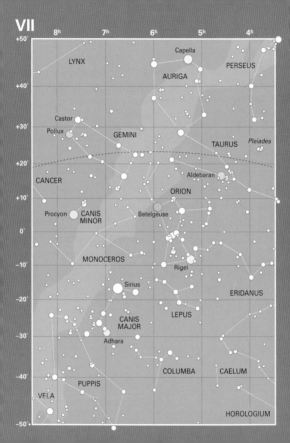

VII

LYNX

PERSEUS

Capella

AURIGA

Castor
Pollux

GEMINI

TAURUS

Pleiades

CANCER

Aldebaran

ORION

Procyon

CANIS
MINOR

Betelgeuse

MONOCEROS

Rigel

ERIDANUS

Sirius

CANIS
MAJOR

LEPUS

Adhara

COLUMBA

CAELUM

PUPPIS

VELA

HOROLOGIUM

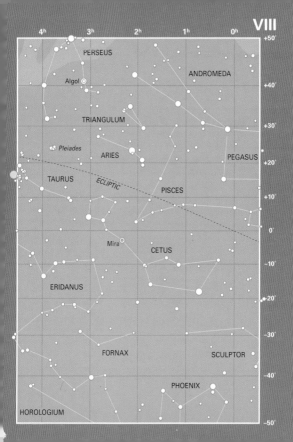

VIII

PERSEUS

Algol

ANDROMEDA

TRIANGULUM

Pleiades

ARIES

PEGASUS

TAURUS

ECLIPTIC

PISCES

Mira

CETUS

ERIDANUS

FORNAX

SCULPTOR

PHOENIX

HOROLOGIUM

IX

CENTAURUS

CRUX
Mimosa
Rigil Kent · Hadar · Acrux
CARINA
LUPUS
CIRCINUS
MUSCA
CHAMAELEON
NORMA
TRIANGULUM AUSTRALE
SCORPIUS
APUS
OCTANS
ARA
-90°
TELESCOPIUM
PAVO
HYDRUS
CORONA AUST.
SAGITTARIUS
INDUS
TUCANA
MIC
PHOENIX
GRUS

12ʰ
13ʰ
14ʰ
15ʰ
16ʰ
17ʰ
18ʰ
19ʰ
20ʰ
21ʰ
22ʰ
23ʰ
0ʰ

-40°
-50°
-60°
-70°
-80°
-70°
-60°
-50°
-40°

Magnitudes
0   1   2

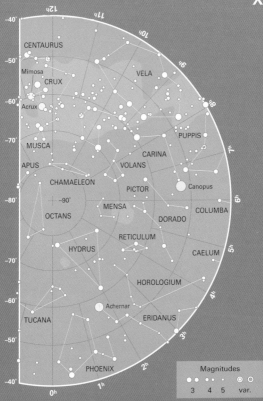

Constellations

ANDROMEDA

A large constellation of the northern hemisphere of the sky that represents the daughter of Queen Cassiopeia and King Cepheus. In Greek mythology, Andromeda was chained to a rock as a sacrifice to the sea monster Cetus, but was saved from certain death by the intervention of the hero Perseus.

BRIGHT STAR

α (alpha) Andromedae (Alpheratz or Sirrah) is a blue-white star of mag. 2.1. Its two alternative names both come from the Arabic meaning 'horse's navel', for it was once considered to be part of neighbouring Pegasus. It now marks the head of Andromeda.

PLANETARY SYSTEM

υ (upsilon) Andromedae, a 4th-mag. star 44 l.y. away, is the first star around which a system of several planets has been detected by professional astronomers.

DOUBLE STARS

γ (gamma) Andromedae is a glorious double star. Even small telescopes easily divide its two components of mags. 2.2 and 4.8. Their beautifully contrasting colours of yellow and blue make this one of the showpiece doubles of the sky.

▶

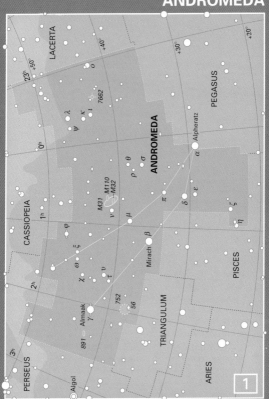

56 Andromedae is a pair of unrelated 6th-mag. orange giant stars, divisible in binoculars, near the star cluster NGC 752.

OPEN CLUSTER
NGC 752 is a large, scattered binocular cluster of about 60 stars of 9th and 10th mags.

GALAXIES
M31 (NGC 224), the Great Andromeda Galaxy, is one of the most famous objects in the entire sky, and the most distant object visible to the naked eye. It is a spiral galaxy similar to our own Milky Way, lying at least 2.4 million l.y. away. On clear, dark nights it can be seen with the naked eye as a hazy smudge; binoculars show its highly elongated shape and rounded core. Only the central part of the galaxy is visible through amateur instruments. Long-exposure photographs reveal its full extent, spanning nearly 3° of sky, i.e. six times the width of the full Moon. Also visible in small amateur telescopes is a satellite galaxy, known both as M32 and NGC 221, which appears like a fuzzy 8th-mag. star ½° south of M31's core. Another satellite galaxy, known alternatively as M110 or NGC 205, is larger but fainter, and lies over 1° north-west of M31.

PLANETARY NEBULA
NGC 7662 is one of the easiest planetary nebulae to see with small amateur telescopes. Under high magnification it appears as a fuzzy blue-green disk with an elliptical outline.

The Andromeda Galaxy, M31, with its companions M32 (below centre) and M110 (upper right). Bill Schoening, Vanessa Harvey/REU program/AURA/NOAO/NSF

ANTLIA The Air Pump

A faint and obscure constellation in the southern hemisphere of the sky, representing the pump invented by the French physicist Denis Papin. There are no bright stars in Antlia.

DOUBLE STAR
ζ^1 ζ^2 (zeta^1 zeta^2) Antliae is a wide pair of 6th-mag. stars easily seen in binoculars. Small telescopes show that ζ^1 Antliae also has a 7th-mag. companion.

APUS The Bird of Paradise

An unremarkable constellation in the south polar region of the sky. It contains no bright stars.

DOUBLE STAR
δ^1 δ^2 (delta^1 delta^2) Apodis is a pair of 5th-mag. orange giant stars easily visible in binoculars.

ANTLIA

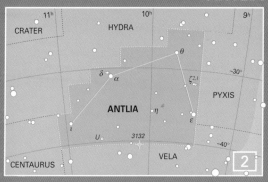

CRATER
HYDRA
11ʰ   10ʰ   9ʰ
θ
δ
α
ζ²,¹
~-30°
ι
η
PYXIS
ANTLIA
ε
U   3132
-40°
CENTAURUS   VELA

2

APUS

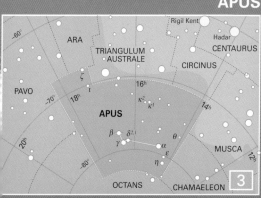

Rigil Kent
-60°
ARA
Hadar
TRIANGULUM AUSTRALE
CENTAURUS
CIRCINUS
PAVO
ζ
ι
16ʰ
-70°
18ʰ
κ²
κ¹
14ʰ
APUS
θ
β
δ²,¹
α
20ʰ   γ   ε
MUSCA
η
12ʰ
-80°
OCTANS   CHAMAELEON

3

AQUARIUS The Water Carrier

A constellation of the zodiac, through which the Sun passes from late February to mid-March. Aquarius represents a man pouring water from a jar.

DOUBLE STAR
ζ (zeta) Aquarii is the central star of the Y-shaped grouping that makes up the water jar of Aquarius. It is a tight pair of 4th-mag. white stars, needing telescopes of at least 75 mm aperture and high magnification to be divided.

GLOBULAR CLUSTERS
M2 (NGC 7089) is a rich globular cluster of 6th mag. It appears as a hazy patch in binoculars and small telescopes.

M72 (NGC 6981) is a 9th-mag. globular cluster, unimpressive in small telescopes.

PLANETARY NEBULAE
NGC 7009, the Saturn Nebula, appears as a small greenish disk of 8th mag. in small telescopes. It is named the Saturn Nebula because in large telescopes its shape resembles that of the planet Saturn.

NGC 7293, the Helix Nebula, is the closest planetary nebula to us, about 300 l.y. away, and the largest in apparent size – its diameter is $\frac{1}{4}°$, half that of the

▶

AQUARIUS

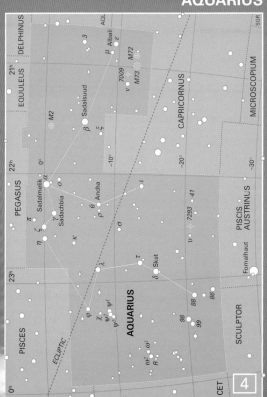

Moon. Although large it is faint, and is best seen in binoculars, or telescopes with low magnification, which show it as a circular misty patch. Its full beauty is brought out in long-exposure photographs, where it appears like two overlapping loops of gas.

METEORS
Three meteor showers radiate from Aquarius each year. The richest shower, the η (eta) Aquarids, reaches a maximum of about 35 per hour around May 5; the δ (delta) Aquarids produce about 20 meteors per hour around July 29; and on August 6 the ι (iota) Aquarids reach about 8 meteors per hour.

AQUILA The Eagle

A constellation straddling the celestial equator, representing an eagle.

BRIGHT STAR
α (alpha) Aquilae (Altair, Arabic for 'flying eagle') is a white star of mag. 0.8. It forms one corner of the so-called Summer Triangle of stars, the other two stars being Deneb in Cygnus and Vega in Lyra. Altair is 17 l.y. away, one of the closest stars to us.

VARIABLE STARS
η (eta) Aquilae is one of the brightest Cepheid variables, ranging from mag. 3.6 to mag. 4.5 every 7.2 days.

▶

AQUILA

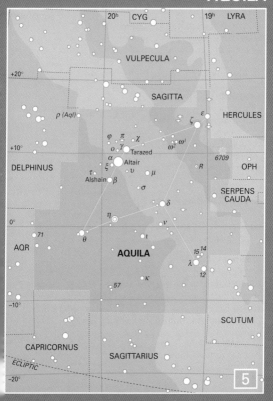

R Aquilae is a red giant variable of the same type as Mira which comes into binocular range every 9 months or so, reaching mag. 5.5. at its best.

DOUBLE STARS
15 Aquilae is a 5th-mag. orange giant with a 7th-mag. companion easily visible in small telescopes.

57 Aquilae is an easy pair of 6th-mag. stars for small telescopes.

ARA The Altar

A small constellation of the southern celestial hemisphere, visualized by the Greeks as the altar on which the gods of Olympus swore an oath of allegiance. It has no stars brighter than 3rd mag.

OPEN CLUSTER
NGC 6193 is a binocular cluster of about 30 stars, the brightest of which is 6th mag.

GLOBULAR CLUSTER
NGC 6397 is a relatively close globular cluster, 10,500 l.y. away. In binoculars and small telescopes it appears as a 6th-mag. smudge about half the size of the Moon.

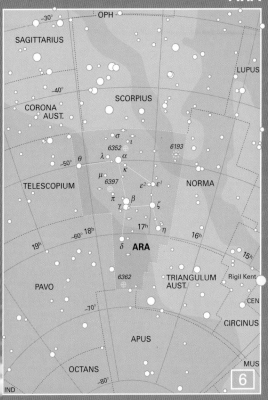

ARA

OPH

SAGITTARIUS

SCORPIUS

LUPUS

CORONA
AUST.

NORMA

TELESCOPIUM

σ
ι
6352
α
λ
μ
6397
π
κ
ε²   ε¹
β
γ
ζ
η

θ

17ʰ

16ʰ

18ʰ

19ʰ

δ   ARA

15ʰ

Rigil Kent

PAVO

6362

TRIANGULUM
AUST.

CEN

CIRCINUS

APUS

OCTANS

MUS

IND

6193

−30°

−40°

−50°

−60°

−70°

−80°

6

ARIES The Ram

An important constellation of the zodiac; the Sun is in Aries from late April to mid-May. Aries represents the ram whose golden fleece was sought by Jason and the Argonauts. The importance of Aries lies in the fact that it once contained the vernal equinox, i.e. the point where the Sun crosses the celestial equator moving from south to north; this is the zero point of right ascension, the celestial equivalent of the Greenwich meridian. The effect of precession has now moved the vernal equinox into neighbouring Pisces, but for historical reasons the vernal equinox is still referred to as the First Point of Aries.

BRIGHT STAR
α (alpha) Arietis (Hamal, from the Arabic for 'lamb'), is a yellow giant of mag. 2.0.

DOUBLE STARS
γ (gamma) Arietis is an easy pair for small telescopes, consisting of two white stars of 5th mag.

λ (lambda) Arietis is a 5th-mag. white star with a 7th-mag. companion visible in binoculars or small telescopes.

π (pi) Arietis, a blue-white 5th-mag. star, has a close 8th-mag. companion visible in small telescopes under high magnification.

ARIES

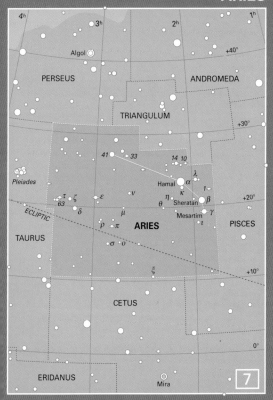

AURIGA The Charioteer

A large constellation of the northern sky, representing a chariot-driver. The shape of the constellation is completed by the star once known as γ (gamma) Aurigae, which is now allocated to Taurus and is dealt with under that constellation as β (beta) Tauri.

BRIGHT STAR

α (alpha) Aurigae (Capella), mag. 0.1, is the 6th-brightest star in the entire sky. It is a close pair of yellow giants, 42 l.y. away.

VARIABLE STARS

ε (epsilon) Aurigae is an eclipsing binary of exceptionally long period. Every 27 years it sinks from mag. 2.9 to 3.8 as it is eclipsed by a dark companion; the eclipses last for a full year. Its next eclipse will start in 2009.

RT Aurigae is a Cepheid variable that rises and falls between mags. 5.0 and 5.8 every 3.7 days.

DOUBLE STARS

θ (theta) Aurigae is a tight double of mag. 2.6 with a 7th-mag. companion requiring at least 100 mm aperture and high magnification to distinguish.

4 Aurigae is a double star for small telescopes consisting of stars of 5th and 8th mags.

▶

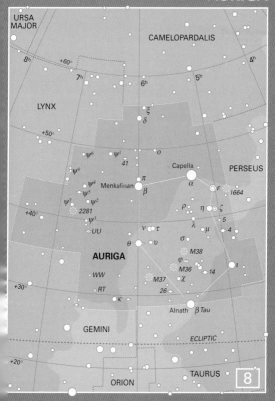

URSA
MAJOR

CAMELOPARDALIS

8ʰ   +60°

7ʰ   6ʰ   5ʰ   4ʰ

LYNX

+50°

ξ
δ

ο

ψ⁶   ψⁱ
41

ψ⁹   π   Capella   PERSEUS
ψ⁴   Menkalinan   α   ε   1664
ψ⁵   ψ²   β
+40°   ψ⁷ 2281   ρ   η   ζ
ψ¹   λ   μ   5
UU   ν τ   σ   4
θ υ   ι
AURIGA   M38
φ
M36
WW   χ   14
M37
+30°   26
RT
κ   Alnath β Tau

GEMINI   ECLIPTIC

+20°

ORION   TAURUS

8

OPEN CLUSTERS

M36 (NGC 1960) is a binocular cluster of 60 stars.

M37 (NGC 2099) is the richest of the clusters in Auriga, containing about 150 stars.

M38 (NGC 1912) is a large, scattered binocular cluster of about 100 stars. South of it lies NGC 1907, a much smaller and fainter cluster.

BOÖTES The Herdsman

A large constellation of the northern celestial hemisphere, representing a man herding a bear (the adjoining constellation of Ursa Major).

BRIGHT STAR

α (alpha) Boötis (Arcturus, from the Greek for 'bear-keeper'), mag. −0.1, is the 4th-brightest star in the entire sky. It is an orange giant 37 l.y. away.

DOUBLE STARS

ε (epsilon) Boötis (Izar or Pulcherrima) is a glorious but difficult pair with contrasting colours of orange and blue, of 3rd and 5th mags. Their closeness requires at least 75 mm aperture and high power to split.

μ (mu) Boötis, a blue-white star of 4th mag., has a 7th-mag. binocular companion. The companion is itself ▶

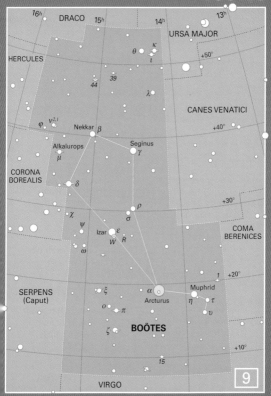

9

double, as revealed under high power by telescopes of at least 75 mm aperture.

ν^1 ν^2 (nu^1 nu^2) Boötis are a wide pair of unrelated orange and white 5th-mag. stars.

ξ (xi) Boötis is a beautiful duo for small telescopes, consisting of a yellow and an orange star of 5th and 7th mags.

METEORS
The Quadrantids, the year's most abundant meteor shower, radiate from the northern part of Boötes, reaching a maximum of about 100 meteors per hour on January 3–4. The shower takes its name from a now-abandoned constellation, Quadrans Muralis, the Mural Quadrant, that once occupied this area of sky. Although they are plentiful, the Quadrantids are not as bright as other great showers such as the Perseids.

CAELUM The Chisel

A faint and easily overlooked constellation of the southern sky. Its brightest star is of mag. 4.4.

DOUBLE STAR
γ (gamma) Caeli is a pair of 5th- and 8th-mag. stars for small telescopes.

CAELUM

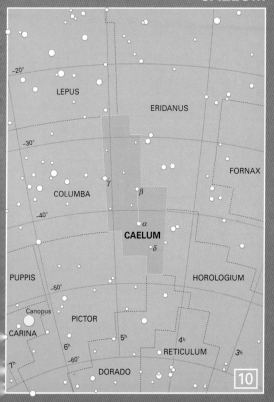

CAMELOPARDALIS The Giraffe

A large but inconspicuous constellation in the north polar region of the sky, which also used to be called Camelopardus. Its brightest stars are of 4th magnitude.

DOUBLE STARS

β (beta) Camelopardalis is a yellow supergiant of mag. 4.0, the brightest star in the constellation; it has a wide 7th-mag. companion visible in small telescopes and good binoculars.

Struve 1694 (Σ 1694), also known as 32 Cam, is an attractive pair of 5th- and 6th-mag. white stars easily split in small telescopes.

OPEN CLUSTER

NGC 1502 is a small binocular cluster of about 45 stars. Small telescopes reveal an easy 7th-mag. double star at its centre. Note also the adjoining chain of faint stars called Kemble's Cascade, which extends for 5 Moon diameters.

GALAXY

NGC 2403 is an 8th-mag. spiral galaxy visible in binoculars or small telescopes under good conditions.

CAMELOPARDALIS

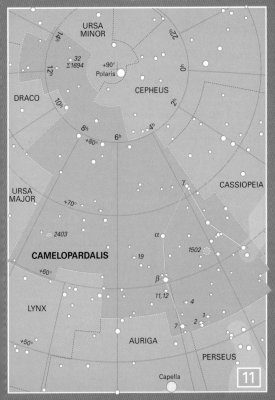

URSA MINOR

14ʰ

32
Σ1694

+90°
Polaris

12ʰ

22ʰ

0ʰ

CEPHEUS

DRACO

10ʰ

2ʰ

8ʰ
+80°

6ʰ

4ʰ

γ

CASSIOPEIA

URSA MAJOR

+70°

α

1502

2403

CAMELOPARDALIS

19

β

+60°

11,12

4

LYNX

1

7

2

+50°

AURIGA

PERSEUS

Capella

11

CANCER The Crab

A constellation of the zodiac, inside whose boundaries the Sun lies from late July to early August. The constellation represents the Crab that attacked Hercules during his fight with Hydra, the Water Snake. In times past, the Sun used to lie in Cancer when at its farthest north of the celestial equator (the northern summer solstice, June 21). Hence the latitude on Earth at which the Sun appears overhead when farthest north came to be known as the Tropic of Cancer. However, because of the effect of precession, the Sun now lies in neighbouring Gemini at the time of the summer solstice.

DOUBLE STARS

ζ (zeta) Cancri is an attractive pair of yellow stars of 5th and 6th mags. visible through small telescopes.

ι (iota) Cancri is a 4th-mag. yellow giant with a 7th-mag. companion easily seen with small telescopes.

OPEN CLUSTER

M44 (NGC 2632), Praesepe ('the manger'), also popularly termed the Beehive Cluster, is a large swarm of about 50 stars of 6th mag. and fainter. To the naked eye it appears as a misty patch. It is best seen in binoculars because of its great size, covering 1½° of sky, three times the apparent width of the full Moon. M44 lies 577 l.y. away.

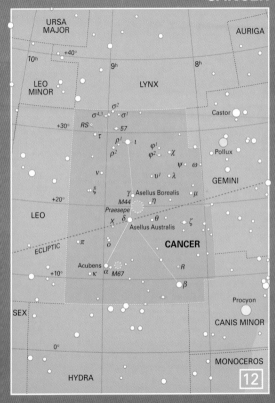

CANCER

URSA
MAJOR

AURIGA

LEO
MINOR

LYNX

10ʰ   +40°

9ʰ   8ʰ

Castor

σ⁴·³   σ²
σ¹

RS   57

τ   ρ¹   ι

ρ²

φ¹

φ²   χ

Pollux

ψ   ω

υ¹   λ

GEMINI

ν

ξ

γ   Asellus Borealis

M44   η

μ

+30°

+20°

Praesepe

χ   δ

θ

ζ

LEO

Asellus Australis

CANCER

π

ο

ECLIPTIC

Acubens

R

κ   α   M67

β

+10°

Procyon

SEX

CANIS MINOR

0°

MONOCEROS

HYDRA

12

CANES VENATICI The Hunting Dogs

A constellation of the northern hemisphere of the sky, representing the two dogs of Boötes, the Herdsman, which it adjoins.

DOUBLE STAR
α (alpha) Canum Venaticorum is known as Cor Caroli, meaning 'Charles's Heart', a name given in honour of the beheaded King Charles I of England. It is a blue-white star of mag. 2.9 with a 6th-mag. companion easily seen in small telescopes.

GLOBULAR CLUSTER
M3 (NGC 5272), of 6th mag., is visible as a hazy patch in binoculars. Although telescopes of 100 mm are needed to resolve individual stars, smaller telescopes show it as a softly glowing ball of light.

GALAXIES
M51 (NGC 5194) is the famous Whirlpool Galaxy, a spiral with a satellite galaxy, NGC 5195, at the end of one of its arms. Large telescopes are needed to show it to effect; in small telescopes it appears as a hazy patch with starlike points marking the nuclei of the main galaxy and its satellite.

M94 (NGC 4736) is an 8th-mag. spiral galaxy seen face on, looking like a small comet in amateur telescopes.

CANES VENATICI

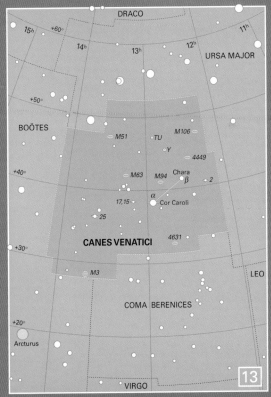

DRACO

15h +60°

14h

13h

12h

11h

URSA MAJOR

+50°

BOÖTES

M51

TU

M106

Y

4449

+40°

M63

M94

Chara

β

2

17,15

α

Cor Caroli

25

CANES VENATICI

4631

+30°

M3

LEO

COMA BERENICES

+20°

Arcturus

VIRGO

13

CANIS MAJOR The Greater Dog

A prominent constellation in the southern hemisphere of the sky, containing many bright stars, notably Sirius, the brightest star of all. It represents one of the dogs of Orion, the other dog being represented by Canis Minor.

BRIGHT STARS

α (alpha) Canis Majoris (Sirius, from the Greek for 'scorching' or 'searing') is the brightest star in the entire sky, of mag. −1.4. It is a white star 8.6 l.y. away, one of the closest stars to the Sun. It has an 8th-mag. white dwarf companion that orbits it every 50 years; this star, Sirius B, is visible only in large amateur telescopes.

β (beta) Canis Majoris (Mirzam, 'the announcer', i.e. of Sirius) is a blue giant of mag. 2.0.

δ (delta) Canis Majoris (Wezen) is a yellow-white super-giant of mag. 1.8, about 1800 l.y. away.

ε (epsilon) Canis Majoris (Adhara), mag. 1.5, is a blue giant with a 7th-mag. companion, difficult to see in small telescopes because of the glare from the main star.

OPEN CLUSTER

M41 (NGC 2287) is a large, bright cluster whose brightest stars are of 7th mag. It is just visible to the naked eye and is well seen in binoculars and small

▶

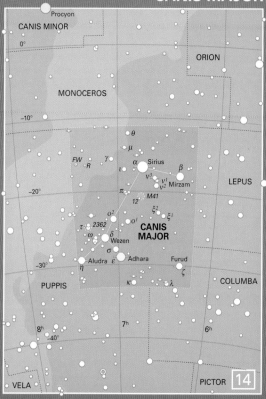

CANIS MINOR

Procyon

0°

ORION

MONOCEROS

−10°

θ

μ

FW

R

γ

α Sirius

β

ι

ν³

ν¹

ν² Mirzam

−20°

π

12

M41

ξ²

ξ¹

o²

o¹

CANIS
MAJOR

τ

2362

ω

δ

Wezen

σ

Aludra

η

ε

Adhara

Furud

ζ

−30°

PUPPIS

κ

λ

COLUMBA

8ʰ

−40°

7ʰ

6ʰ

VELA

PICTOR

LEPUS

14

telescopes. Its brightest stars seem to be arranged in chains.

NGC 2362 is a compact cluster for small telescopes, surrounding the 4th-mag. blue supergiant τ (tau) Canis Majoris.

CANIS MINOR The Lesser Dog

A small constellation lying on the celestial equator. It represents the smaller of the two dogs of Orion, Canis Major being the larger. Apart from Procyon, its brightest star, it contains little of interest. Procyon forms a triangle of brilliant stars with Sirius (in Canis Major) and Betelgeuse (in Orion).

BRIGHT STAR
α (alpha) Canis Minoris (Procyon, from the Greek meaning 'before the dog', i.e. it rises before Canis Major), mag. 0.4, is the eighth-brightest star in the sky. It is a yellow-white star 11.4 l.y. away, one of the nearest stars to the Sun.

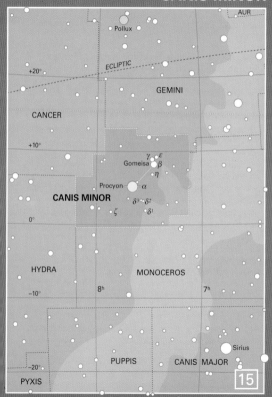

AUR

Pollux

ECLIPTIC

+20°

GEMINI

CANCER

+10°

γ ε
Gomeisa β
η
Procyon α
CANIS MINOR
δ³ δ²
δ¹
0°
ζ

HYDRA

MONOCEROS

-10°

8ʰ

7ʰ

PUPPIS

CANIS MAJOR

Sirius

-20°

PYXIS

15

CAPRICORNUS The Sea Goat

The smallest constellation of the zodiac, through which
the Sun passes from late January to mid-February. It
represents a goat with a fish tail. In ancient times the
Sun lay in Capricornus at the northern winter solstice,
its farthest point south of the equator. The effect of
precession has now moved the winter solstice into
neighbouring Sagittarius, but the latitude on Earth at
which the Sun appears overhead on that day is still
known as the Tropic of Capricorn.

DOUBLE STARS

α (alpha) Capricorni (Algedi or Giedi, from the Arabic
meaning 'the kid'), is a naked-eye pair of unrelated 4th-
mag. stars, one a yellow supergiant and the other an
orange giant. Each star is itself double. Small telescopes
show that α^1 (alpha^1) Capricorni, the fainter of the pair,
has a 9th-mag. companion and α^2 (alpha^2) Capricorni
has an 11th-mag. companion.

β (beta) Capricorni is a 3rd-mag. star with a 6th-mag.
companion for binoculars or small telescopes.

GLOBULAR CLUSTER

M30 (NGC 7099) is an 8th-mag. globular cluster next
to the 5th-mag. star 41 Capricorni.

CAPRICORNUS

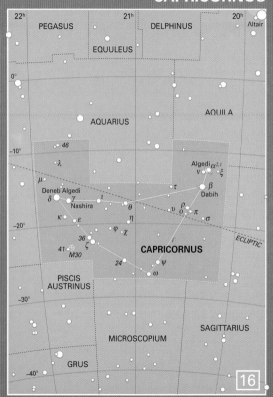

CARINA The Keel

A large constellation of the southern celestial hemisphere. Carina is one of the parts into which the constellation of Argo Navis, the Ship of the Argonauts, was divided. It lies in an area of the Milky Way rich with clusters and nebulae. The stars ι (iota) and ε (epsilon) Carinae, together with κ (kappa) and δ (delta) Velorum, form the so-called False Cross which can be mistaken for the real Southern Cross.

BRIGHT STARS

α (alpha) Carinae (Canopus), mag. −0.6, is the second-brightest star in the sky. It is a white supergiant just over 300 l.y. away.

β (beta) Carinae (Miaplacidus) is a blue-white star of mag. 1.7.

ε (epsilon) Carinae is an orange giant star of mag. 1.9.

VARIABLE STARS

η (eta) Carinae is one of the most erratically behaved stars in the heavens. In 1843 it flared up to mag. −1, then settled around 6th mag. until 1998 when it brightened to 5th mag. It lies inside the nebula NGC 3372 (see p. 89). η Carinae is thought to be either an unstable supergiant star with a mass of over 100 Suns that throws off shells of gas and dust at irregular intervals, or a massive binary. It is expected to become a supernova within the next 10,000 years.

▶

CARINA

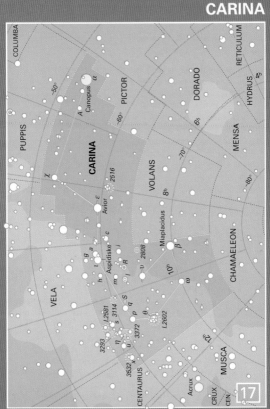

17

l Carinae is a Cepheid variable that fluctuates between mags. 3.3 and 4.2 every 35.5 days.

R Carinae is a red giant variable of the same type as Mira. It has a period of about 10 months and reaches 4th mag. at its brightest.

DOUBLE STAR

υ (upsilon) Carinae is a 3rd-mag. white star with a 6th-mag. companion visible in small telescopes.

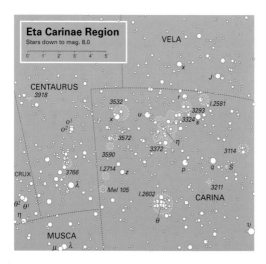

OPEN CLUSTERS

IC 2602 is a large and bright cluster sometimes called the 'southern Pleiades', centred on the 3rd-mag. blue-white star θ (theta) Carinae. It contains six naked-eye stars and many more are visible in binoculars and small telescopes, covering 1° of sky.

NGC 2516 is a large binocular cluster of 80 stars, containing a 5th-mag. red giant and three double stars of 8th and 9th mags. for small telescopes.

NGC 3114 is a binocular cluster the width of the Moon, a poorer version of NGC 2516.

NGC 3532 is a large binocular cluster of 150 stars of 7th mag. and fainter, very rich in small telescopes which show its markedly elongated shape. A 4th-mag. yellow star at one edge is not a member of the cluster, but is a much more distant supergiant.

NEBULA

NGC 3372, the η (eta) Carinae Nebula, is a diffuse nebula visible to the naked eye, larger than the Orion Nebula. It is bisected by a dark V-shaped lane of dust. NGC 3372 contains the erratic variable star η Carinae, which lies in the nebula's bright central part near a dark notch called the Keyhole because of its shape. This is a star-studded region for sweeping with binoculars (see map opposite).

CASSIOPEIA

A distinctive W-shaped constellation of the northern hemisphere of the sky, representing a mythical Queen of Ethiopia, wife of King Cepheus and mother of Andromeda.

VARIABLE STAR
γ (gamma) Cassiopeiae is an unstable blue giant that throws off shells of gas at irregular intervals, causing it to vary unpredictably between 2nd and 3rd mags.

DOUBLE AND MULTIPLE STARS
η (eta) Cassiopeiae is a beautiful double star for small telescopes consisting of yellow and red components of 4th and 7th mags.

ι (iota) Cassiopeiae is a 5th-mag. white star with an 8th-mag. companion divisible in small telescopes. The brighter star has a close 7th-mag. companion visible with 100 mm aperture and high magnification.

OPEN CLUSTERS
M52 (NGC 7654) is a binocular cluster of about 100 stars, somewhat kidney-shaped and with a prominent 8th-mag. orange star at one edge.

NGC 457 is an attractive cluster for small telescopes, including the 5th-mag. white supergiant φ (phi) Cassiopeiae. The stars of NGC 457 are seemingly arranged in chains.

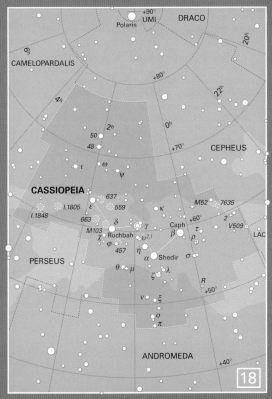

CASSIOPEIA

18

CENTAURUS The Centaur

A resplendent constellation of the southern celestial
hemisphere, representing the mythical beast known as a
centaur. It contains the closest known star to the Sun,
α (alpha) Centauri, which is actually a family of three
stars.

BRIGHT STARS

α (alpha) Centauri (Rigil Kentaurus, from the Arabic
meaning 'centaur's foot') appears to the naked eye as a
star of mag. −0.3, the third-brightest in the sky. Small
telescopes split it into a pair of yellow stars of mags. 0.0
and 1.4 which orbit each other every 80 years. These
two stars lie 4.4 l.y. away from us. About 0.2 l.y. closer
is an 11th-mag. red dwarf called Proxima Centauri; this
is strictly the closest star of all to us. Proxima Centauri
lies 2° away in the sky from its two brighter companions, so it is not even in the same telescopic field of
view.

β (beta) Centauri (Hadar or Agena) is a blue giant of
mag. 0.6, the 11th-brightest star in the sky. A line
drawn from α through β Centauri points to Crux, the
Southern Cross.

DOUBLE STAR

3 Centauri is a neat pair of blue-white stars of 5th and
6th mags. divisible through small telescopes.

▶

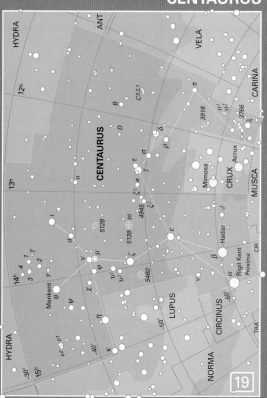

CENTAURUS

HYDRA
ANT
VELA
CARINA

12h

π
3918
o¹
o²
3766
λ

B
C 3.2.1

D

CENTAURUS

ρ
δ
σ
τ
γ
e
ζ¹
ζ²
ξ²
4945
ω
5139
5128
ι
d
n
μ
ν
φ
ζ
5460

Mimosa
Acrux
CRUX
MUSCA

J
Hadar
ε
β
Rigil Kent
Proxima
α
CIR

13h

Menkent
γ
θ
ψ
χ
υ¹
υ²
η
κ

T
1
2
3
4

c²
c¹

HYDRA
14h
15h

-30°
-40°
-50°
-60°

LUPUS
CIRCINUS
NORMA
TRA

19

PLANETARY NEBULA

NGC 3918 is an 8th-mag. planetary nebula, popularly known as the Blue Planetary, with a small blue-green disk similar in appearance to the planet Uranus.

GLOBULAR CLUSTER

ω (omega) Centauri (NGC 5139) is the largest and brightest globular cluster in the sky, covering a larger area than the full Moon. To the naked eye it appears as a fuzzy 4th-mag. star, somewhat elliptical in outline. Small telescopes or even binoculars begin to resolve its outer regions into a granular mass of sparkling stars. It is one of the closer globulars to us, 17,000 l.y. away.

GALAXY

NGC 5128 is a peculiar galaxy also known as the radio source Centaurus A, visible in binoculars or small tele-scopes as a 7th-mag. hazy blur. Long-exposure photo-graphs show that it is a giant elliptical galaxy crossed by a dark band of dust (see photograph opposite). This peculiar object may be the result of an elliptical galaxy merging with a spiral galaxy, the spiral galaxy providing the lane of dust. NGC 5128 is one of the strongest sources known to radio astronomers, who have detected lobes of radio emission either side of the galaxy, as though it has ejected clouds of gas in a series of explosions. It lies about 15 million l.y. away.

The enormous radio galaxy NGC 5128, also known as Centaurus A, can be seen in small telescopes. ESO

CEPHEUS

A constellation of the north polar region of the sky, representing the mythical King of Ethiopia who was the husband of Cassiopeia and father of Andromeda. It contains the celebrated variable star δ (delta) Cephei.

VARIABLE STARS

δ (delta) Cephei is the prototype of the most important type of variable star known to astronomers, the Cepheid variables. Thanks to the so-called Period–Luminosity Law, astronomers can derive a Cepheid's absolute magnitude by observing its period of variation; Cepheid variables are therefore the 'standard candles' by which astronomers measure distances in space. The variability of δ Cephei was discovered in 1784 by the English amateur astronomer John Goodricke. The star is a yellow-white supergiant about 1000 l.y. away that varies between mags. 3.5 and 4.4 every 5 days 9 hours. δ Cephei is also a double star (see page 98).

μ (mu) Cephei is a red supergiant, prototype of a class of semi-regular variables. It varies between mags. 3.4 and 5.1 with no set period. William Herschel named it the Garnet Star because of its pronounced coloration: it is one of the reddest stars visible to the naked eye.

T Cephei is a red giant variable of the same type as Mira, which brightens to 5th or 6th magnitude every 13 months or so.

▶

CEPHEUS

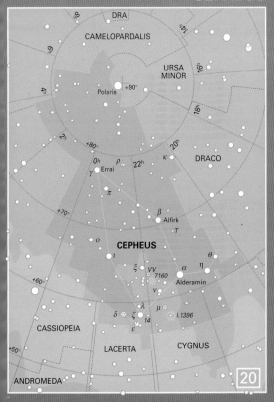

VV Cephei is one of the largest stars known, perhaps 1000 times the diameter of our Sun. It varies semi-regularly between mags. 4.8 and 5.4.

DOUBLE STARS

β (beta) Cephei is a blue giant of 3rd mag. with an 8th-mag. companion visible in small telescopes. β Cephei is also a variable star, although its fluctuations are too small to be noticed with the naked eye, less than 0.1 mag. every 4.6 hours. It is the prototype of a class of pulsating variable stars (also known as β Canis Majoris stars) with periods of a few hours and slight changes in light output.

δ (delta) Cephei, in addition to being a noted variable star (see page 96 and comparison chart opposite), is an attractive double star for small telescopes or even binoculars. The brighter cream-coloured component, which is the variable, is accompanied by a wide 6th-mag. blue-white star.

ξ (xi) Cephei is a double star for small telescopes consisting of a blue-white star of 4th mag. with a 6th-mag. yellowish companion.

o (omicron) Cephei is a 5th-mag. yellow giant with a close 7th-mag. companion visible in apertures of 60 mm and above with high magnification.

Right: Comparison chart (top) and light curve (below) for the variable star Delta Cephei

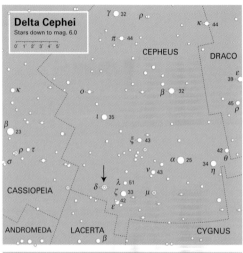

Delta Cephei
Stars down to mag. 6.0

0' 1' 2' 3' 4' 5'

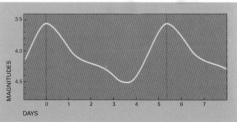

CETUS The Whale

A large constellation straddling the celestial equator,
representing the sea monster that nearly devoured
Andromeda before her rescue by Perseus.

BRIGHT STAR
β Ceti (Deneb Kaitos or Diphda), is a yellow giant of
mag. 2.0, the brightest star in the constellation.

NEARBY STAR
τ (tau) Ceti is a 3rd-mag. yellow dwarf, 11.9 l.y. away.
Of all the nearby single stars, this one is most like our
own Sun.

VARIABLE STAR
o (omicron) Ceti (Mira, 'the amazing one'), a red giant,
is the prototype of the long-period variables. Mira fluctuates between about 3rd and 9th mags. with an average period of 330 days. Its variability was first noticed
in 1596 by the Dutch astronomer David Fabricius.

DOUBLE STARS
α (alpha) Ceti (Menkar, from the Arabic for 'nose') is a
red giant of 3rd mag., forming a binocular double with
a more distant 6th-mag. blue star.

γ (gamma) Ceti is a close pair of 4th- and 6th-mag. stars
requiring at least 60 mm aperture and high power to
split.

▶

CETUS

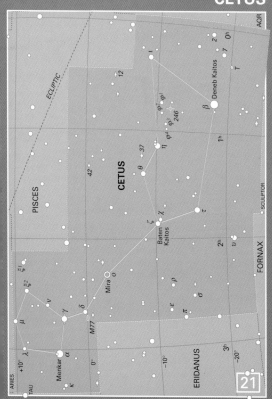

AQR

ECLIPTIC

PISCES

CETUS

Deneb Kaitos

β

φ² φ¹
246
φ⁴
η
θ
37
42

χ
ζ
Baten Kaitos

τ

ν

SCULPTOR

FORNAX

Mira
ο

ρ

ξ¹
ξ²
ν
γ
δ

ε
π
σ

M77
α
Menkar
λ
κ

μ

ARIES
TAU

+10°

0°

–10°

ERIDANUS

3ʰ

0ʰ

1ʰ

2ʰ

–20°

21

GALAXY

M77 (NGC 1068) is a 9th-mag. spiral galaxy presented face-on. It is unimpressive in small telescopes, but its main interest is that it is the brightest of the so-called Seyfert galaxies, spiral galaxies with bright centres that are closely related to quasars (intensely luminous objects far off in the Universe).

CHAMAELEON The Chameleon

An insignificant constellation in the south polar region of the sky, representing a chameleon. Its brightest stars are of 4th mag.

DOUBLE STAR

δ (delta) Chamaeleontis is a binocular pair of blue and orange stars of 4th and 5th mags. They each lie at a similar distance from us, roughly 360 l.y., but are too widely separated to form a true binary.

PLANETARY NEBULA

NGC 3195 is a faint planetary nebula, appearing of similar size to the planet Jupiter through a telescope.

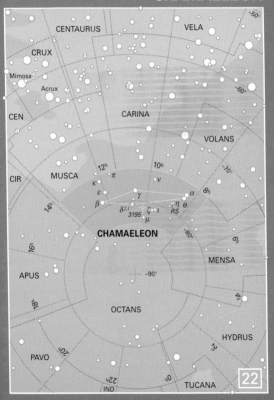

CENTAURUS
VELA
CRUX
Mimosa
Acrux
CEN
CARINA
VOLANS
CIR
MUSCA
14ʰ
12ʰ
κ
π
ε
β
10ʰ
ν
γ
δ²·¹
3195
μ
ζ
ι
η
θ
RS
α
8ʰ
CHAMAELEON
–50°
–60°
–70°
–80°
16ʰ
APUS
18ʰ
–90°
MENSA
6ʰ
OCTANS
4ʰ
2ʰ
PAVO
20ʰ
22ʰ
IND
0ʰ
TUCANA
HYDRUS
22

CIRCINUS The Compasses

A small and obscure constellation of the southern
hemisphere of the sky, overshadowed by neighbouring
Centaurus.

DOUBLE STAR

α (alpha) Circini is a 3rd-mag. white star with a wide
9th-mag. companion visible in small telescopes.

COLUMBA The Dove

A constellation of the southern hemisphere of the sky,
representing the dove that followed Noah's Ark.
It contains no objects of particular interest to amateur
observers.

RUNAWAY STAR

μ (mu) Columbae, a 5th-mag. blue star, is one of three
so-called runaway stars that are diverging from the
Orion Nebula region at speeds of around 100 km/sec.
The other two runaways are 53 Arietis and AE Aurigae.
μ Columbae and AE Aurigae are almost identical stars,
but moving in opposite directions. They are thought to
have been members of a binary which split apart after
encountering another binary in the Orion Nebula about
2.5 million years ago. The other binary involved in the
encounter may have been ι (iota) Orionis.

CIRCINUS

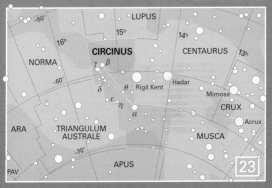

COLUMBA

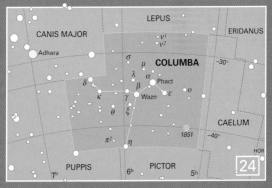

COMA BERENICES Berenice's Hair

A faint constellation of the northern sky, representing the locks of the Egyptian Queen Berenice which she offered to the gods for the return of her husband from battle. Coma Berenices contains a number of galaxies within range of amateur telescopes; these are members of the Virgo Cluster (see diagram page 247).

DOUBLE STAR
24 Comae Berenices, a 5th-mag. orange giant, makes a beautiful colour contrast with a 7th-mag. blue-white companion, visible in small telescopes.

OPEN CLUSTER
The Coma Star Cluster (also known as Melotte 111) consists of a scattered group of about 50 stars representing Queen Berenice's severed tresses. They form a V-shaped binocular group to the south of 4th-mag. γ (gamma) Comae Berenices.

GLOBULAR CLUSTER
M53 (NGC 5024) is an 8th-mag. globular cluster, visible as a misty patch in small telescopes.

GALAXY
M64 (NGC 4826) is a 9th-mag. spiral galaxy visible in small telescopes. It is popularly termed the Black Eye Galaxy because of a dark patch of dust near its centre, shown by telescopes of 150 mm and larger.

COMA BERENICES

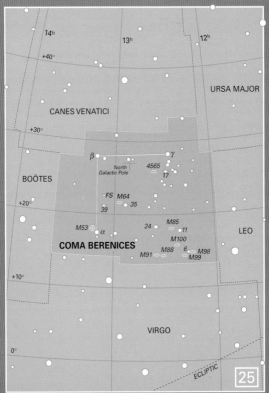

14ʰ
13ʰ
12ʰ
+40°

URSA MAJOR

CANES VENATICI

+30°

β
North
Galactic Pole
γ
4565
17

BOÖTES

FS
M64
35

+20°

39

M53
α
24
M85
11
M100
M91
M88
6
M98
M99

COMA BERENICES

LEO

+10°

VIRGO

0°

ECLIPTIC

25

CORONA AUSTRALIS The Southern Crown

A small constellation of the southern hemisphere of the sky, lying between Sagittarius and Scorpius. Corona Australis has been known since Greek times and is easily identified by its distinctive circlet shape, even though its brightest stars are of only 4th mag. Despite its small size it is not without interest, lying on the edge of the Milky Way.

DOUBLE STARS

γ (gamma) Coronae Australis is a tight pair of 5th-mag. yellow stars requiring at least 100 mm aperture and high magnification to split.

κ (kappa) Coronae Australis is a wide pair of 6th-mag. blue-white stars easily divided by small telescopes.

λ (lambda) Coronae Australis is a 5th-mag. blue-white star with a wide 10th-mag. companion visible in small telescopes.

GLOBULAR CLUSTER

NGC 6541 is a 7th-mag. globular cluster, 22,000 l.y. away, for binoculars and small telescopes.

CORONA AUSTRALIS

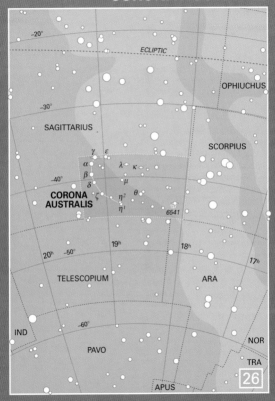

CORONA BOREALIS The Northern Crown

A small constellation of the northern hemisphere of the sky. Its most distinctive feature is a crescent-shaped group of stars in which the constellation's brightest star, mag. 2.2 Alphekka (also known as Gemma), is set like a jewel in the crown.

VARIABLE STARS

R Coronae Borealis is a celebrated variable star given to catastrophic changes in brightness at intervals of a few years. Normally it is of 6th mag. but it can unpredictably drop in a few weeks to as low as 15th mag., subsequently taking some months to return to its former brightness. Its variation is thought to be due to sooty carbon particles in its outer layers, which periodically build up and then blow away.

T Coronae Borealis, known as the Blaze Star, is a recurrent nova that normally hovers around 11th mag. but which in 1866 and 1946 brightened to 2nd and 3rd mags. respectively. Further eruptions may be expected at any time.

DOUBLE STARS

ζ (zeta) Coronae Borealis is a pair of 5th- and 6th-mag. blue-white stars for small telescopes.

$\nu^1 \nu^2$ (nu^1 nu^2) Coronae Borealis is a wide binocular pair of 5th-mag. red and orange giants.

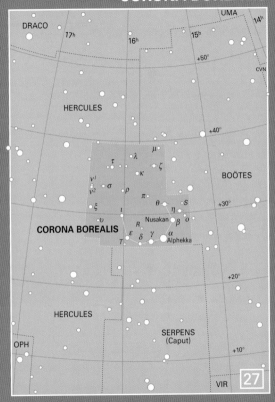

27

CORVUS The Crow

A small constellation of the southern hemisphere of the sky. In Greek legends, Corvus is linked with Crater and Hydra. The crow (Corvus) is sent on an errand to fetch water in a cup (Crater), but returns with the water snake (Hydra) instead.

DOUBLE STARS

δ (delta) Corvi is an unequal double for small telescopes, consisting of a 3rd-mag. blue-white star accompanied by a wide 9th-mag. companion.

Struve 1669 (Σ 1669) is a matching pair of 6th-mag. white stars divisible by small telescopes.

CRATER The Cup

An insignificant constellation representing the goblet of Apollo, and associated in mythology with neighbouring Corvus and Hydra. Its brightest stars are of 4th mag. and there are no objects of particular interest for amateur astronomers.

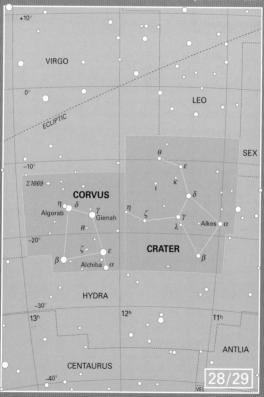

CRUX The Southern Cross

The smallest constellation in the sky, but one of the most famous and distinctive. It lies in a rich region of the Milky Way, on the borders of Centaurus.

BRIGHT STARS

α (alpha) Crucis (Acrux) is of mag. 0.8 to the naked eye and is the 13th-brightest star in the sky. But small telescopes split it into two sparkling blue-white components of mags. 1.3 and 1.7.

β (beta) Crucis (Mimosa) is a blue-white giant star of mag. 1.2.

DOUBLE STAR

μ (mu) Crucis is an easy pair of blue-white stars of 4th and 5th mags. for small telescopes.

OPEN CLUSTER

NGC 4755 is a showpiece cluster known as the Jewel Box because of its glittering appearance in telescopes. Its brightest star is the 6th-mag. blue supergiant κ (kappa) Crucis; the whole cluster is sometimes termed the Kappa Crucis Cluster.

NEBULA

The Coalsack is a dark cloud of dust that blots out the bright star background of the Milky Way. The entire Coalsack measures 7° × 5° and extends into neighbouring Centaurus and Musca.

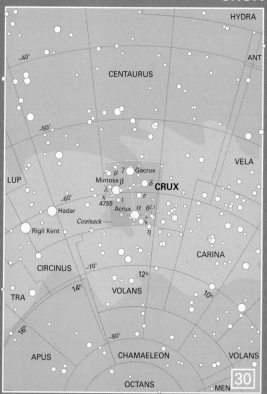

CRUX

HYDRA

CENTAURUS

ANT

VELA

-40°

-50°

LUP

Mimosa β

μ γ

Gacrux

δ **CRUX**

ε

λ

κ

ι

4755

Acrux α θ² ι

Hadar

-60°

-70°

Rigil Kent

Coalsack

η

CARINA

CIRCINUS

12h

TRA

14h

VOLANS

10h

16h

-80°

APUS

CHAMAELEON

VOLANS

OCTANS

MEN

30

CYGNUS The Swan

A prominent constellation of the northern sky, sometimes known as the Northern Cross because of its cruciform shape. Cygnus represents a swan flying along the Milky Way. In legend Cygnus is the Greek god Zeus in disguise. The Milky Way is particularly rich in the region of Cygnus.

BRIGHT STAR
α (alpha) Cygni (Deneb, from the Arabic meaning 'tail') is a white supergiant star of mag. 1.2, lying over 3,000 light years away; it is the most distant first-magnitude star.

VARIABLE STARS
χ (chi) Cygni is a red giant long-period variable which reaches its brightest, up to 3rd mag., every 400 days. At its faintest it sinks to 14th mag.

P Cygni is a blue supergiant of 5th mag. which has flared up to 3rd mag. in the past, apparently as a result of throwing off shells of gas.

DOUBLE STARS
β (beta) Cygni (Albireo) is a beautiful coloured double star, one of the showpieces of the sky. It consists of an orange giant of 3rd mag. accompanied by a 5th-mag. blue-green star visible in small telescopes or even good binoculars. Not to be missed by any observer. ▶

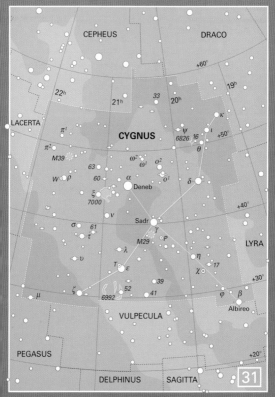

CYGNUS

CEPHEUS   DRACO

22h   21h   33   20h   +60°

19h

LACERTA

π¹   ψ   κ   +50°
6826   16   ι
π²   θ
M39   **CYGNUS**
63   ω²   o²   θ
60   ω¹   o¹
W   ρ   α   δ
ξ   Deneb   +40°
7000
ν
σ   61   Sadr
τ   γ   LYRA
λ   M29   P
υ   η   17
τ   ε   χ
ζ   39   φ   β
μ   52   41   +30°
6992   Albireo

VULPECULA   +20°

PEGASUS   DELPHINUS   SAGITTA   31

o¹ (omicron¹) Cygni is a 4th-mag. orange giant that forms an attractive wide binocular duo with the 5th-mag. blue-white giant 30 Cygni. Closer to o¹ Cygni is a 7th-mag. blue companion visible in binoculars or small telescopes. o¹ Cygni is an eclipsing binary with a very long period of nearly 10.5 years, varying by less than two-tenths of a magnitude.

61 Cygni is a famous pair of orange dwarf stars of 5th and 6th mags., easily split in small telescopes. They lie 11.4 l.y. away and are among the closest neighbours to the Sun.

PLANETARY NEBULA
NGC 6826 is known as the Blinking Planetary because it seems to blink on and off as one looks alternately at it and away from it. Telescopes of 75 mm aperture show its 8th-mag. pale blue disk. The nebula lies within 1° of 6th-mag. 16 Cygni, an easy double star for small telescopes.

NEBULAE
NGC 6992 is the brightest part of the loop of gas that makes up the Veil Nebula; it can be seen in wide-angle telescopes with low power or (under the best conditions) with binoculars. The Veil Nebula itself is the remains of a supernova that exploded some 5000 years ago and consists of a number of faint streamers in a loop shape. Most of the loop is seen well only on long-exposure photographs.

The Cygnus Rift, also known as the Northern Coalsack, is a dark lane of dust that divides the bright Milky Way in this region, as can be seen by the naked eye on clear nights.

NGC 7000 is a large patch of bright nebulosity known as the North America Nebula because it is shaped like that continent. It can be glimpsed by the naked eye as a brightening in the Milky Way with a fish-hook extension. Despite its size it is a difficult object for amateur instruments because of its low surface brightness and is best seen in binoculars on clear dark nights.

BLACK HOLE
Although by definition invisible, the object known as Cygnus X-1 is thought to be the first black hole ever identified. Its existence is known because it is an X-ray source, orbiting a 9th-mag. blue supergiant known as HDE 226868. It lies one Moon diameter from the star η (eta) Cygni.

RADIO GALAXY
Cygnus A, near γ (gamma) Cygni but vastly more distant, is one of the strongest radio sources known to astronomers. It is a radio galaxy of 15th mag., beyond the reach of all but large telescopes. Long-exposure photographs show Cygnus A as a pair of fuzzy blobs in contact, evidently two galaxies merging after a collision.

DELPHINUS The Dolphin

A small constellation in the equatorial region of the sky
with a distinctive kite-shape. It represents a dolphin,
one of the messengers of the sea-god Poseidon.
Although not bright, its four main stars form a box-
shaped group known as Job's Coffin. The peculiar
names of two of these stars, Sualocin and Rotanev, first
appeared in a star catalogue published at Palermo
Observatory, Italy, in 1814. Read backwards, they give
Nicolaus Venator, which is the Latinized form of
Niccolò Cacciatore, the observatory's assistant director
at that time. Delphinus lies on the edge of the Milky
Way in an area noted for novae.

DOUBLE STARS

γ (gamma) Delphini is a noted double star consisting of
golden and yellow-white stars of 4th and 5th mags.,
divisible in small telescopes. Struve 2725, a fainter and
closer double of 8th mag stars, lies nearby in the same
field of view.

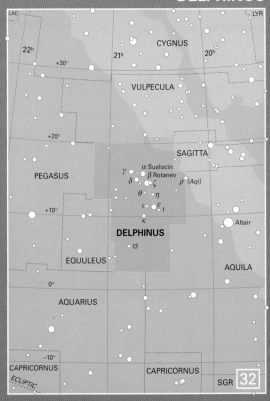

LAC

LYR

22h

CYGNUS

21h

20h

+30°

VULPECULA

+20°

SAGITTA

γ α Sualocin
δ   β Rotanev
  θ   ζ   ρ (Aql)
     η
  ι   ε₁
  κ

PEGASUS

+10°

DELPHINUS

Altair

13

EQUULEUS

AQUILA

0°

AQUARIUS

−10°

CAPRICORNUS

CAPRICORNUS

ECLIPTIC

SGR

32

DORADO The Goldfish

A constellation of the southern sky, sometimes also known as the Swordfish. Its main feature is the Large Magellanic Cloud, the larger and closer of the two satellite galaxies that accompany our Milky Way.

VARIABLE STAR
β (beta) Doradus is one of the brightest Cepheids, varying from mags. 3.5 to 4.1 every 9 days 20 hours.

GALAXY
The Large Magellanic Cloud is an irregularly shaped galaxy about 170,000 l.y. away, containing perhaps 10,000 million stars (less than one-tenth the number in our own Galaxy). It is visible to the naked eye as a hazy patch 6° across, like a detached portion of the Milky Way. Binoculars and small telescopes show that it is richly studded with bright stars, clusters and nebulae.

NEBULA
NGC 2070 is a glowing cloud of hydrogen gas in the Large Magellanic Cloud, visible to the naked eye as a fuzzy star. It is popularly known as the Tarantula Nebula because of its spidery shape. At the centre of the Tarantula is a cluster of supergiant stars, which includes some of the most massive stars known. Were it as close as the Orion Nebula it would fill the entire constellation of Orion and cast shadows.

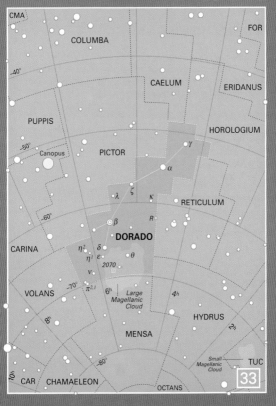

CMA
COLUMBA
FOR
−40°
CAELUM
ERIDANUS
HOROLOGIUM
PUPPIS
−50°
Canopus
PICTOR
γ
α
λ ζ κ
RETICULUM
−60°
β
R
DORADO
CARINA
η² δ
η¹ ε θ
ι
2070
ν
π²·¹ 6ʰ Large Magellanic Cloud 4ʰ
−70°
VOLANS
8ʰ
HYDRUS
2ʰ
−80°
MENSA
Small Magellanic Cloud
TUC
10ʰ
CAR CHAMAELEON
OCTANS

33

DRACO The Dragon

An extensive constellation of the northern polar region. It represents the dragon that guarded the golden apples in the garden of the Hesperides, and lies with one of the feet of Hercules, its slayer, firmly planted upon its head. Its brightest star is γ (gamma) Draconis, mag. 2.2, named Etamin from the Arabic meaning 'the serpent'.

DOUBLE AND MULTIPLE STARS

μ (mu) Draconis is a close pair of 6th-mag. stars, divisible with high magnification through telescopes of 75 mm aperture.

ν (nu) Draconis is an outstanding binocular pair of 5th-mag. white stars.

o (omicron) Draconis is a 5th-mag. yellow giant with an 8th-mag. companion visible in small telescopes.

ψ (psi) Draconis is a pair of 5th- and 6th-mag. yellow-white stars divisible in small telescopes or even through binoculars.

16–17 Draconis is a binocular pair of 5th- and 6th-mag. blue-white stars. Telescopes show a 7th-mag. companion closer to the brighter star of the pair, making a striking triple system.

39 Draconis is another impressive triple star. Binoculars show it as a wide pair of blue-white and yellow 5th- ▶

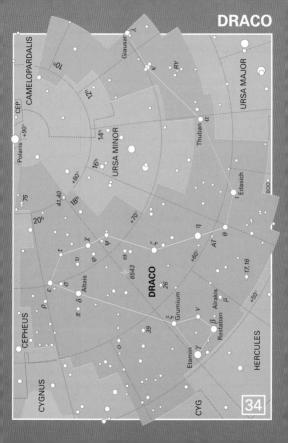

DRACO

CAMELOPARDALIS

URSA MAJOR

γ

Giausar

κ

RY

10ʰ

12ʰ

CEP.

Polaris +90°

14ʰ

α

Thuban

URSA MINOR

16ʰ

+80°

ι Edasich

18ʰ

BOO

75

41,40

+70°

θ

20ʰ

τ

χ

ν

ψ

φ

ω

ζ

η

AT

+60°

17,16

σ

ε

6543

DRACO

26

+50°

ρ

π

δ

Altais

μ

ν

Alrakis

CEPHEUS

39

ξ

Grumium

β

Rastaban

HERCULES

ο

γ

Etamin

CYGNUS

CYG

34

and 7th-mag. stars, but telescopes reveal an 8th-mag. companion closer to the brighter star.

40–41 Draconis is an easy pair of 6th-mag. yellow stars for small telescopes.

PLANETARY NEBULA
NGC 6543 is one of the most prominent planetary nebulae, appearing as an irregularly shaped blue-green disk of 9th mag., like an out-of-focus star, in small telescopes.

EQUULEUS The Little Horse

The second-smallest constellation, lying in the equatorial region of the sky next to Pegasus, the larger horse. It is sometimes also called the Foal or Colt. Its brightest star is of 4th mag.

DOUBLE AND MULTIPLE STARS
γ (gamma) Equulei is a 5th-mag. white star with a 6th-mag. binocular companion, 6 Equulei, which is unrelated.

1 Equulei, also known as ε (epsilon) Equulei, is a 5th- and 7th-mag. yellow-white pair divisible in small telescopes. The brighter star is itself double but the two components, which orbit every 101 years, are currently too close to be separated in amateur telescopes.

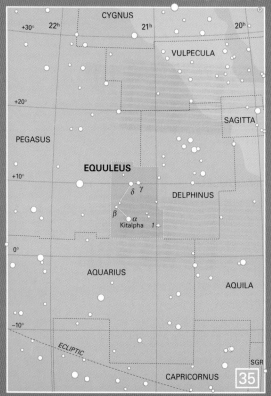

CYGNUS

VULPECULA

SAGITTA

PEGASUS

EQUULEUS

δ γ

β

α
Kitalpha 1

DELPHINUS

AQUARIUS

AQUILA

ECLIPTIC

CAPRICORNUS

SGR

35

ERIDANUS The River

An extensive constellation, sixth-largest in the sky, that meanders from the celestial equator far into the southern celestial hemisphere. In Greek legends Eridanus represents the river into which Phaethon fell after his disastrous attempt to drive the chariot of his father the Sun God, but it has also been identified with real rivers such as the Po of Italy, the Nile, and the Euphrates. Despite its size Eridanus has few bright stars. It contains a number of distant galaxies which are too faint for amateur telescopes, but which show up well on long-exposure photographs.

BRIGHT STAR
α (alpha) Eridani (Achernar, from the Arabic meaning 'end of the river') is a blue-white star of mag. 0.5, the 9th-brightest star in the entire sky.

NEARBY STAR
ε (epsilon) Eridani, mag. 3.7, is a yellow dwarf similar to the Sun, lying 10.5 l.y. away. It has a planet of similar size to Jupiter.

DOUBLE AND MULTIPLE STARS
θ (theta) Eridani is an impressive pair of 3rd- and 4th-mag. blue-white stars for small telescopes.

o² (omicron²) Eridani (also known as 40 Eridani) is a remarkable triple star containing the most easily

▶

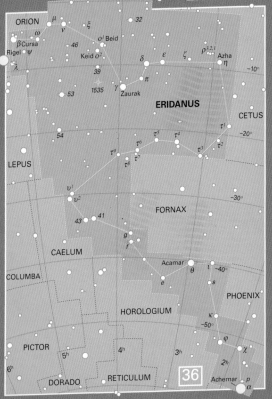

ERIDANUS

ORION

μ
ξ
32
ν
ο¹ Beid
ω
46
β Cursa
Keid ο²
δ
ε
ζ
ρ³·²·¹
Azha
Rigel
ψ
39
π
η
λ
γ
-10°
53
1535
Zaurak
ERIDANUS

CETUS

54
τ⁵ τ⁴
τ¹
-20°
τ⁶
τ³
τ⁹
τ¹ τ²
LEPUS
τ⁸ τ⁷

υ¹
-30°
υ²

43 41
FORNAX

CAELUM
g
f

COLUMBA
Acamar
ι
-40°
θ
s
e
PHOENIX

HOROLOGIUM
κ
-50°

PICTOR
φ
5ʰ 4ʰ
3ʰ
χ
6ʰ
2ʰ
DORADO
RETICULUM
36
Achernar ρ
α

observable white dwarf in the sky. The 4th-mag. main star is a Sun-like yellow dwarf. Small telescopes show a 10th-mag. companion; this is the white dwarf. Small telescopes also reveal an 11th-mag. third member of the system. This is a red dwarf, completing a rare trio that is not to be missed.

32 Eridani is a beautiful coloured double for small telescopes, consisting of 5th- and 6th-mag. stars of yellow and blue-green colour.

PLANETARY NEBULA
NGC 1535 is a 9th-mag. planetary nebula visible as a bluish disk in small telescopes.

FORNAX The Furnace

A barren constellation of the southern hemisphere of the sky, originally known as Fornax Chemica (the chemical furnace). It contains a number of faint galaxies, beyond the reach of all but large amateur telescopes, which are its main interest.

DOUBLE STAR
α (alpha) Fornacis is a 4th-mag. yellow star with a close 7th-mag. companion, possibly variable, visible in small telescopes.

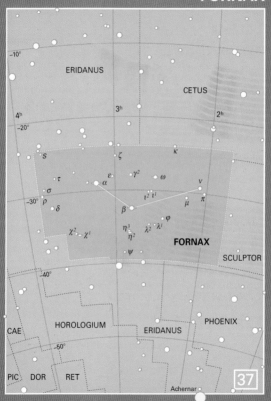

FORNAX

ERIDANUS

CETUS

FORNAX

SCULPTOR

HOROLOGIUM

ERIDANUS

PHOENIX

CAE

PIC   DOR   RET

Achernar

37

GEMINI The Twins

Gemini is a major zodiacal figure, the Sun passing through the constellation from late June to late July.

BRIGHT STARS

α (alpha) Geminorum (Castor) appears of mag. 1.6 to the naked eye. Small telescopes with high magnification split it into two sparkling white stars of mags. 1.9 and 2.9. Also visible is a wider 9th-mag. red dwarf. All three stars are spectroscopic binaries, so Castor actually consists of six stars, all lying 52 l.y. away.

β (beta) Geminorum (Pollux) is an orange giant of mag. 1.2, 34 l.y. away.

VARIABLE STARS

ζ (zeta) Geminorum is a 4th-mag. Cepheid variable that fluctuates by 0.5 mag. every 10.2 days.

η (eta) Geminorum is a red giant semi-regular variable, ranging from 3rd to 4th mag. every 230 days or so.

DOUBLE STAR

38 Geminorum is an attractive white and yellow pair of 5th- and 8th-mag. stars for small telescopes.

OPEN CLUSTER

M35 (NGC 2168) is an outstanding cluster of over 100 stars, just resolvable in binoculars; telescopes show its stars to be arranged in curving chains.

▶

GEMINI

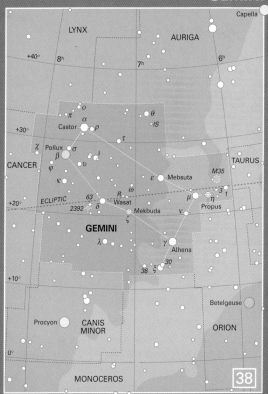

38

PLANETARY NEBULA
NGC 2392 is shown by small telescopes as an 8th-mag.
blue-green oval about the size of the disk of Saturn.
It is popularly known as the Eskimo or Clown Face
Nebula from its appearance through large telescopes.

METEORS
The Geminid meteors, one of the year's most prominent showers, radiate from near Castor around 13–14
December each year. At maximum as many as 100
meteors per hour can be seen, many of them bright and
explosive.

GRUS The Crane

A constellation of the southern hemisphere of the sky,
representing a water bird, the crane.

BRIGHT STAR
α (alpha) Gruis (Alnair, 'the bright one'), is a blue-
white star of mag. 1.7.

DOUBLE STARS
δ (delta) Gruis is a naked-eye double of unrelated 4th-
mag. stars.

μ (mu) Gruis is a naked-eye double of unrelated 5th-
mag. stars.

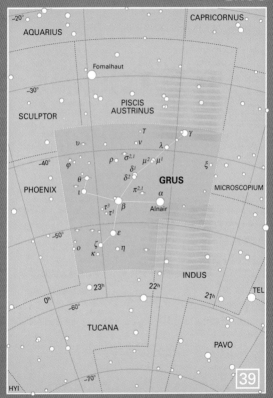

GRUS

AQUARIUS

CAPRICORNUS

Fomalhaut

SCULPTOR

PISCIS
AUSTRINUS

υ

τ
ν
λ
γ

φ

ρ
$\sigma^{2,1}$
μ^2 μ^1
ξ

PHOENIX

θ
ι

δ^1
δ^2
$\pi^{2,1}$

GRUS

MICROSCOPIUM

α
Alnair

β

τ^3
τ^1

ε

o
ζ
κ
η

INDUS

23^h
22^h
21^h

0^h

TUCANA

PAVO

HYI

39

TEL

HERCULES

A northern hemisphere constellation, the fifth-largest in
the sky. In Greek legend, Hercules was the son of Zeus
who was set 12 labours in punishment for killing his
wife and children in a fit of madness. A group of four
stars makes up a shape known as the Keystone, which
marks the pelvis of Hercules; on one side of the
Keystone lies the celebrated globular cluster M13, the
finest object of its kind in the northern sky.

VARIABLE STARS

α (alpha) Herculis (Rasalgethi, from the Arabic
meaning 'the kneeler's head') is a red giant that varies
erratically between 3rd and 4th mag. Small telescopes
show that it is also a double star, with a 5th-mag. blue-
green companion.

68 Herculis is an eclipsing binary that varies between
mags. 4.7 and 5.4 every 2 days.

DOUBLE STARS

κ (kappa) Herculis is a 5th-mag. yellow giant with an
unrelated 6th-mag. companion, easily seen in small
telescopes.

ρ (rho) Herculis is a pair of 5th-mag. blue-white stars
divisible in small telescopes.

95 Herculis is an attractive pair of 5th-mag. stars for
small telescopes, appearing gold and silver.

▶

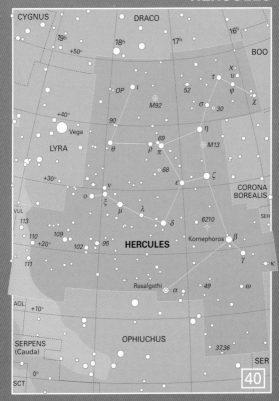

HERCULES

CYGNUS

DRACO

BOO

19ʰ

18ʰ

17ʰ

16ʰ

+50°

OP

ι

χ

υ

τ

52

φ

σ

30

χ

M92

90

+40°

Vega

η

M13

θ

ρ

69

π

LYRA

68

+30°

ε

ζ

CORONA
BOREALIS

ν

ο

ξ

λ

SER

μ

δ

6210

VUL

113

Kornephoros

β

110

109

HERCULES

102

95

γ

κ

111

Rasalgethi

α

49

ω

AQL

+10°

SERPENS
(Cauda)

OPHIUCHUS

37,36

SCT

0°

SER

40

GLOBULAR CLUSTERS

M13 (NGC 6205) contains 300,000 stars and is the brightest globular cluster in the northern skies, a show-piece for all apertures. It is visible to the naked eye as a 6th-mag. misty patch on clear nights, and is prominent in binoculars. Small telescopes begin to resolve the cluster into stars, giving it the appearance of a mottled mound of sparkling stardust. M13 lies 25,000 l.y. away.

M92 (NGC 6341) is a smaller, somewhat fainter and more distant cluster than M13, but readily visible in binoculars. It is more condensed at the centre than M13 and at first glance may appear starlike.

HOROLOGIUM The Pendulum Clock

A faint and barren constellation of the southern skies, containing scarcely any objects of note. Its brightest star is of 4th mag.

VARIABLE STARS

R Horologii is a red giant long-period variable that fluctuates over a wide range, 5th to 14th mag., approximately every 400 days.

TW Horologii is a deep-red semi-regular variable that fluctuates between mags. 5.5 and 6.0 every 160 days or so.

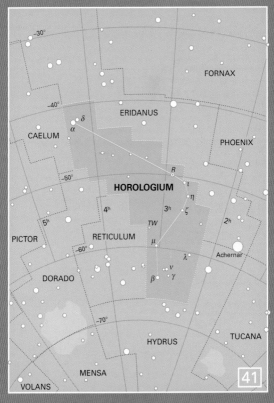

HYDRA The Water Snake

The largest constellation in the sky, yet far from prominent. Its most readily recognizable feature is a group of stars that make up its head, lying just north of the celestial equator. From there its tail snakes away towards Centaurus and Lupus in the south; the total length of the constellation is over 100°. In Greek mythology, Hydra was the multi-headed monster slain by Hercules as one of his 12 labours. In another legend it is linked with the story of Corvus, the crow, which was sent to fetch water in a cup, represented by the constellation Crater. The constellations of Corvus and Crater are found on Hydra's back.

BRIGHT STAR

α (alpha) Hydrae (Alphard, from the Arabic meaning 'the solitary one') is an orange giant of mag. 2.0. It is the only star in the entire constellation that is brighter than mag. 3.0.

VARIABLE STARS

R Hydrae is a red giant long-period variable similar to Mira in Cetus. It fluctuates between 4th and 11th mags. every 390 days.

U Hydrae is a deep-red variable star that fluctuates between 4th and 6th mags. with a period of 115 days.

DOUBLE STARS

ε (epsilon) Hydrae is a challenging double star, consisting of yellow and blue components of 3rd and 7th mags. that require at least 75 mm aperture and high magnification to be separated.

27 Hydrae is a 5th-mag. yellow-white star with a very wide 7th-mag. binocular companion. Small telescopes reveal that this star itself has a 10th-mag. companion.

54 Hydrae is a 5th-mag. yellow-white star with a 7th-mag. purple companion for small telescopes.

OPEN CLUSTER

M48 (NGC 2548) is a large binocular cluster of about 80 stars of 8th mag. and fainter, arranged in a triangular shape. Small telescopes resolve the brightest of its individual stars.

PLANETARY NEBULA

NGC 3242 is a 9th-mag. planetary nebula for small telescopes. It appears as a hazy blue-green disk known as the Ghost of Jupiter because of its similarity to the telescopic appearance of that planet.

GLOBULAR CLUSTER

M68 (NGC 4590) is an 8th-mag. globular cluster, unspectacular in small instruments. Its individual stars are resolvable with apertures of 100 mm and above. ▶

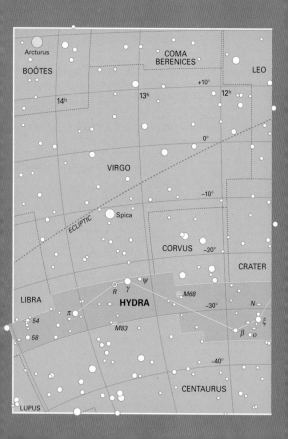

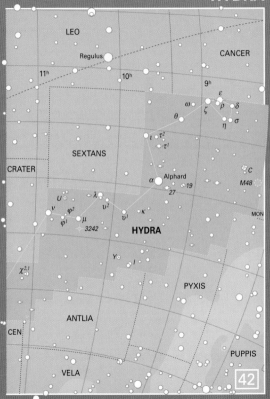

HYDRA

LEO

CANCER

Regulus

11ʰ   10ʰ   9ʰ

ω   ε
θ   ζ   ρ   δ
η   σ
ι   τ²
τ¹

SEXTANS

CRATER

α   Alphard
19
27   C
M48

U   λ
ν   φ²   υ²
φ¹   μ   υ¹   κ   MON
3242   HYDRA

γ

χ²·¹

PYXIS

ANTLIA

CEN

VELA   PUPPIS

42

GALAXY

M83 (NGC 5236) is an impressive 8th-mag. spiral galaxy seen face-on, one of the brightest galaxies in the entire sky. Through small telescopes it appears as a misty patch with a bright nucleus; apertures of 150 mm are needed to trace its spiral arms.

Spiral galaxy M83. Bill Schoening/AURA/NOAO/NSF

HYDRUS The Lesser Water Snake

An insignificant constellation of the south polar region of the sky. Its brightest stars are of 3rd mag.

DOUBLE STAR

π (pi) Hydri is a wide binocular or naked-eye pair of unrelated 6th-mag. red and orange giant stars.

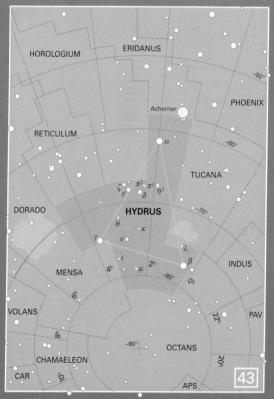

INDUS The Indian

A faint constellation of the southern hemisphere of the sky. It represents a North American native Indian. Its brightest star is of 3rd mag.

NEARBY STAR
ε (epsilon) Indi is an orange dwarf star somewhat cooler than the Sun, 11.8 l.y. away, appearing of mag. 4.7.

DOUBLE STAR
θ (theta) Indi is a pair of 4th- and 7th-mag. stars that can be separated by small telescopes.

LACERTA The Lizard

A very inconspicuous constellation of the northern hemisphere of the sky, sandwiched between Andromeda and Cygnus. Its brightest stars are of 4th mag. Lacerta lies in the Milky Way and has been the site of three naked-eye novae since 1910, but it contains no objects of note for users of small telescopes.

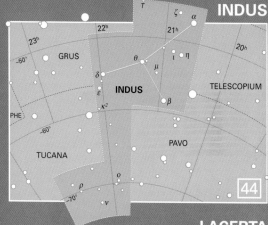

INDUS

23h · 22h · 21h · 20h

−50°

GRUS

θ

ζ · α

T

μ · ι · η

δ

INDUS

ε

κ²

β

TELESCOPIUM

PHE

−60°

TUCANA

PAVO

ρ · o

−70°

ν

44

LACERTA

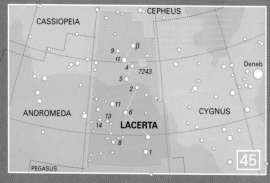

CEPHEUS

CASSIOPEIA

β

9 · α

Deneb

4 · 7243

5

2

ANDROMEDA

11

13

6

14

LACERTA

CYGNUS

8

1

PEGASUS

45

LEO The Lion

A constellation of the zodiac, through which the Sun passes from mid-August to mid-September. It represents the lion slain by Hercules as part of his 12 labours.

BRIGHT STAR

α (alpha) Leonis (Regulus, 'the little king') is a blue-white star of mag. 1.4. Binoculars and small telescopes show a wide 8th-mag. companion.

VARIABLE STAR

R Leonis is a red giant long-period variable that fluctuates between 4th and 11th mags. with an average period of about 310 days.

DOUBLE AND MULTIPLE STARS

γ (gamma) Leonis (Algieba, 'the forehead') is a showpiece pair of yellow giants of 2nd and 4th mags., divisible in small telescopes. In binoculars an unrelated 5th-mag. star, 40 Leonis, is seen nearby.

ζ (zeta) Leonis is an optical triple star of 3rd mag. In binoculars ζ Leonis appears to have two companions of 6th mag. at different distances from it, but both are unrelated to it.

τ (tau) Leonis is a 5th-mag. yellow giant with an 8th-mag. companion for binoculars and small telescopes. ▶

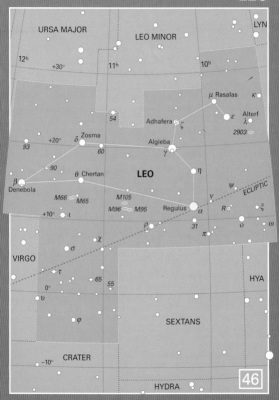

LEO

GALAXIES

Leo contains a number of interesting galaxies, none particularly easy for small telescopes. M65 (NGC 3623) and M66 (NGC 3627) are a pair of 9th-mag. spiral galaxies; M66 is presented at an angle to us and appears cigar-shaped. M95 (NGC 3351) and M96 (NGC 3368) are two more spiral galaxies, of 10th and 9th mags. respectively. All these objects require low power and a dark night to be picked out as faint smudges.

METEORS

The yearly Leonid meteor shower appears around November 17, radiating from a point near γ (gamma) Leonis. Usually the numbers are low, peaking at about 10 per hour, but occasionally tremendous storms of up to 100,000 meteors per hour have been seen. These storms occur when the Earth passes near to the parent comet, Tempel–Tuttle, which has an orbital period of 33 years. The last major storm was in 1966; increased activity was seen in the years 1998 to 2002, but did not reach the extreme levels of 1966.

LEO MINOR The Lesser Lion

A faint and obscure constellation of northern skies, sandwiched between Ursa Major and Leo. It contains no objects for users of small telescopes.

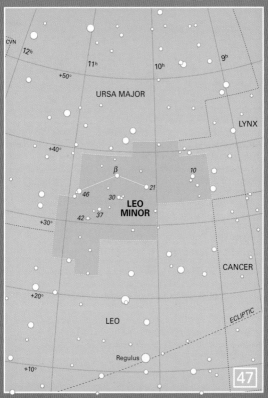

CVN

12ʰ

11ʰ

10ʰ

9ʰ

+50°

URSA MAJOR

LYNX

+40°

β

10

21

46

30

LEO
MINOR

42   37

+30°

CANCER

+20°

ECLIPTIC

LEO

Regulus

+10°

47

LEPUS The Hare

A constellation of the southern hemisphere of the sky. Although overshadowed by neighbouring Orion and Canis Major it is not without interest for amateur observers.

VARIABLE STAR
R Leporis is a deep-red long-period variable, popularly known as Hind's Crimson Star. It varies between 6th and 12th mag. about every 430 days.

DOUBLE STAR
γ (gamma) Leporis is an attractive binocular duo of yellow and orange stars of 4th and 6th mags.

OPEN CLUSTER
NGC 2017 is a star cluster that is really a multiple star. Binoculars and small telescopes show a group of five stars of 6th to 10th mags. Two of these stars are themselves close doubles, as revealed by larger apertures, making this a group of at least seven stars.

GLOBULAR CLUSTER
M79 (NGC 1904) is a compact 8th-mag. globular cluster for small telescopes. It lies near a 5th-mag. triple star, Herschel 3752, also visible in small telescopes.

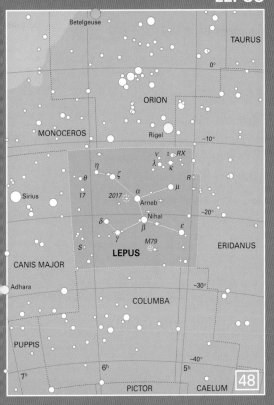

Betelgeuse

TAURUS

0°

ORION

Rigel

MONOCEROS

−10°

ν   ι RX
λ   κ
η   α   R
θ   ξ   μ
        α
17   2017   Arneb
Sirius   Nihal

δ   −20°

β   ε

γ   M79

LEPUS   ERIDANUS

CANIS MAJOR

S

Adhara   −30°

COLUMBA

PUPPIS

−40°

7ʰ   6ʰ   5ʰ

48

PICTOR   CAELUM

LIBRA The Scales

A faint constellation of the zodiac through which the Sun passes during November. It is now visualized as the scales of the goddess of justice, but once represented the claws of neighbouring Scorpius.

VARIABLE STAR

δ (delta) Librae is an eclipsing binary of the same type as Algol. It varies between mags. 4.9 and 5.9 every 2 days 8 hours.

DOUBLE AND MULTIPLE STARS

α (alpha) Librae (Zubenelgenubi, from the Arabic meaning 'southern claw') is a 3rd-mag. blue-white star with a wide 5th-mag. companion easily visible in binoculars.

ι (iota) Librae is a complex multiple star. To the naked eye it appears of 4th mag. Binoculars show a wide 6th-mag. companion, 25 Librae. In small telescopes another, fainter companion is visible; apertures of 75 mm show that this companion actually consists of a close pair of 10th-mag. stars.

μ (mu) Librae is a 6th-mag. star with a close 7th-mag. companion for telescopes of 75 mm aperture.

GLOBULAR CLUSTER

NGC 5897 is a 9th-mag. globular cluster with loosely scattered stars, unspectacular in small instruments.

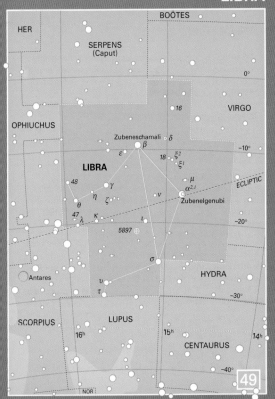

LIBRA

BOÖTES

HER

SERPENS
(Caput)

0°

16

VIRGO

OPHIUCHUS

Zubeneschamali δ

-10°

LIBRA

β

ε

18 ξ²
ξ¹

μ

48

γ

ν

α²·¹

ECLIPTIC

η ζ

Zubenelgenubi

θ

47 κ

ι

λ

5897 ⊕

-20°

σ

HYDRA

Antares

ν

-30°

τ

SCORPIUS

LUPUS

16ʰ

15ʰ

CENTAURUS

14ʰ

-40°

NOR

49

LUPUS The Wolf

A constellation of the southern sky, depicted as a wolf held in the grasp of neighbouring Centaurus.

DOUBLE AND MULTIPLE STARS

ε (epsilon) Lupi is a 3rd-mag. blue-white star with a 9th-mag. companion for small telescopes.

η (eta) Lupi is a 3rd-mag. blue-white star with an 8th-mag. companion, not easy to see in the smallest telescopes because of the brightness difference.

κ (kappa) Lupi is an easy pair of 4th- and 6th-mag. stars for small telescopes.

μ (mu) Lupi is a 4th-mag. star with a 7th-mag. companion visible in small telescopes. With apertures of at least 100 mm the primary is actually seen to be a tight pair of 5th-mag. stars.

ξ (xi) Lupi is a neat pair of 5th- and 6th-mag. stars for small telescopes.

π (pi) Lupi consists of a close pair of 5th-mag. stars, divisible with apertures of 75 mm and above.

OPEN CLUSTER

NGC 5822 is a large 7th-mag. open cluster of 150 faint stars for binoculars and small telescopes.

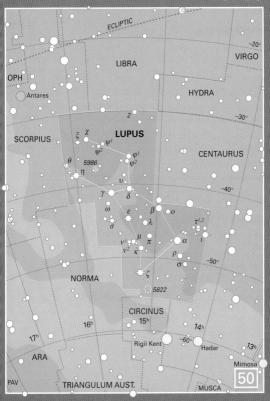

ECLIPTIC

−20°

VIRGO

LIBRA

OPH

HYDRA

Antares

2

−30°

ξ χ

SCORPIUS

LUPUS

ψ¹ ψ¹ φ¹

θ

φ²

CENTAURUS

5986

η

υ

−40°

γ δ

ω ε β ο

d λ τ¹·²

ν¹ μ π α ι

ν² κ ρ

σ −50°

NORMA

ζ

5822

CIRCINUS

15ʰ

14ʰ

13ʰ

16ʰ

−60°

Rigil Kent

Hadar

17ʰ

Mimosa

ARA

50

PAV

TRIANGULUM AUST.

MUSCA

LYNX The Lynx

An exceedingly faint constellation of northern skies, named by the Polish astronomer Johannes Hevelius because only the lynx-eyed would be able to see it. Changes of constellation boundaries have produced confusion in the stellar nomenclature of this region. For instance, Lynx contains the star 10 Ursae Majoris, while the star known as 41 Lyncis lies in Ursa Major.

DOUBLE AND MULTIPLE STARS

5 Lyncis is a 5th-mag. orange giant with an 8th-mag. companion visible in small telescopes.

12 Lyncis is a 5th-mag. blue-white star with a 7th-mag. companion for small telescopes. In apertures of 75 mm or more the brighter component is seen to have a much closer 6th-mag. companion, making 12 Lyncis a triple.

19 Lyncis is a neat pair of 6th- and 7th-mag. stars for small telescopes, with a wider 8th-mag. companion.

38 Lyncis is a tight pair of 4th- and 6th-mag. stars that provide a challenge for small telescopes because of their closeness.

41 Lyncis is to be found over the border in Ursa Major. Small telescopes show it to be an interesting triple star, consisting of components of 5th, 8th and 10th mags., arranged in a triangle.

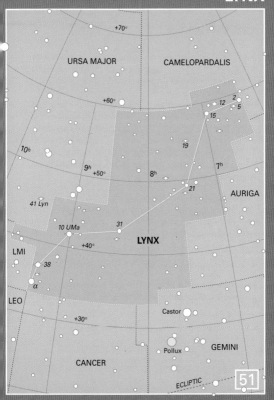

+70°

URSA MAJOR

CAMELOPARDALIS

+60°

12   2
5

15

19

10ʰ

9ʰ +50°

8ʰ

7ʰ

21

AURIGA

41 Lyn

31

10 UMa

LYNX

+40°

LMI

38

α

LEO

+30°

Castor

CANCER

Pollux

GEMINI

ECLIPTIC

51

LYRA The Lyre

A prominent constellation of the northern hemisphere of the sky, representing the harp of Orpheus. Lyra is sometimes also visualized as an eagle or vulture.

BRIGHT STAR

α (alpha) Lyrae (Vega, from the Arabic meaning 'swooping eagle') is a brilliant white star of mag. 0.0, the fifth-brightest in the entire sky. It lies 25 l.y. away. Vega forms one corner of the Summer Triangle of three bright stars, completed by Deneb in Cygnus and Altair in Aquila.

VARIABLE AND DOUBLE STARS

β (beta) Lyrae is an eclipsing cream-coloured binary star that varies from mag. 3.3 to 4.4 every 12 days 22 hours. (For a comparison chart, see page 163). Small telescopes resolve β Lyrae as an attractive double star, with a blue companion of 7th mag. In addition, there are two wider 10th-mag. companions that can be seen in small telescopes.

δ (delta) Lyrae is another double–variable. Binoculars show a 6th-mag. blue-white star and a 4th-mag. red giant semi-regular variable, which varies by about 0.1 mag.

ε (epsilon) Lyrae is the most celebrated quadruple star in the sky, commonly known as the Double Double. ▶

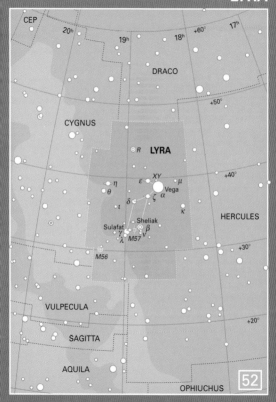

LYRA

Binoculars, or even keen eyesight, show it to consist of a wide pair of 5th-mag. white stars. But each star is itself a close double, requiring at least 60 mm or 75 mm aperture and high magnification to be split.

ζ (zeta) Lyrae is an easy double star of 4th and 6th mags. for binoculars or small telescopes.

R Lyrae is a red giant that varies semi-regularly between mags. 3.9 and 5.0 every 6 or 7 weeks.

PLANETARY NEBULA

M57 (NGC 6720) is the famous Ring Nebula, beautifully shown on long-exposure photographs (see opposite) but somewhat disappointing in amateur telescopes. It is easy to find, midway between β (beta) and γ (gamma) Lyrae. Small telescopes show it as a ghostly elliptical disk on dark nights, larger than the apparent size of Jupiter. Its brightness is similar to that of a 9th-mag. star out of focus. Apertures of at least 150 mm are needed to show its central hole, but its faint central star is beyond the reach of amateur telescopes.

METEORS

The Lyrid meteors radiate from a point near Vega every year. They are bright but not particularly plentiful, peaking at about 10 meteors per hour on April 21–22, although occasional higher bursts of activity have been seen.

Right: Comparison chart for the variable star Beta Lyrae

Above: The Ring Nebula in Lyra, M57. Daniel Folha and Simon Tulloch, Isaac Newton Group of Telescopes, La Palma

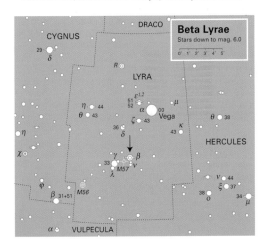

MENSA The Table Mountain

A faint and obscure constellation of the south polar region of the sky. Its main point of interest is that part of the Large Magellanic Cloud strays into it over the border from Dorado. The brightest stars of Mensa are 5th mag. and there are no objects of note for owners of small telescopes.

MICROSCOPIUM The Microscope

A faint constellation of the southern hemisphere of the sky. Its brightest stars are of 5th mag. and it contains scarcely any objects of interest for owners of small telescopes.

DOUBLE STAR
α (alpha) Microscopii is a 5th-mag. yellow giant with a 10th-mag. companion visible in small telescopes.

MENSA

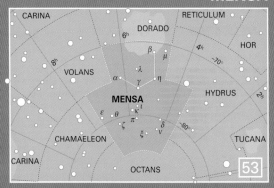

MICROSCOPIUM

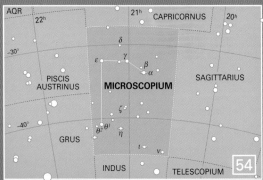

MONOCEROS The Unicorn

A constellation of the equatorial region of the sky. Its
brightest stars are of only 4th mag. and inevitably it is
overshadowed by the brilliance of neighbouring Orion,
but it contains several interesting clusters and nebulae.

VARIABLE STAR
S Monocerotis is an erratically variable blue-white
supergiant which fluctuates around mag. 4.7 with no
set period. It has an 8th-mag. companion, difficult to
see in the smallest telescopes because of its closeness.
S Monocerotis is the brightest member of the star
cluster NGC 2264 (see page 168).

DOUBLE AND MULTIPLE STARS
β (beta) Monocerotis is an outstanding triple star for
telescopes of all apertures. The three blue-white stars,
all of 5th mag., are arranged in an arc, the two faintest
stars being closest together.

δ (delta) Monocerotis, a blue-white star of 4th mag.,
has a wide but unrelated 5th-mag. companion,
21 Monocerotis, visible to the naked eye or binoculars.

8 Monocerotis, also known as ε (epsilon) Monocerotis,
is an attractive double for small telescopes, consisting
of blue and yellow stars of 4th and 7th mags.

▶

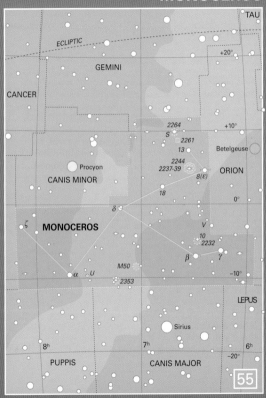

MONOCEROS

TAU

ECLIPTIC

+20°

GEMINI

CANCER

2264
S

2261

13

Betelgeuse

2244
2237-39

ORION

8(ε)

Procyon

CANIS MINOR

18

0°

δ

V

MONOCEROS

10
2232

ζ

β

γ

α U

M50

2353

−10°

LEPUS

Sirius

8ʰ

7ʰ

6ʰ

−20°

PUPPIS

CANIS MAJOR

55

OPEN CLUSTERS

M50 (NGC 2323) is a 6th-mag. binocular cluster.
Telescopes resolve dozens of individual stars including
an orange giant near its southern edge.

NGC 2232 is a scattered binocular cluster centred on
the 5th-mag. blue-white star 10 Monocerotis.

NGC 2244 is a binocular cluster of young stars born
from a faint surrounding nebulosity, the Rosette
Nebula, over 5000 l.y. away. The brightest stars of the
cluster form a noticeable rectangle. A 6th-mag. yellow
star, 12 Monocerotis, appears to be part of the cluster
but is actually an unrelated foreground star. The nebula
surrounding this cluster is known as the Rosette
because of its flower-like shape. The Rosette Nebula
can be glimpsed in binoculars under dark skies, but its
full beauty shows up only on long-exposure photographs (see facing page).

NGC 2264 is another cluster embedded in a nebula.
Binoculars show a group of about 20 stars, including
the variable S Monocerotis (see page 166) arranged in a
triangular shape. Long-exposure photographs show
faint surrounding nebulosity. At its southern end is an
intruding wedge of dark dust, known as the Cone
Nebula because of its tapered shape. NGC 2264 lies
about 2500 l.y. away.

The Rosette Nebula is a faint loop of gas surrounding the star cluster NGC 2244 in Monoceros. AURA/NOAO/NSF.

MUSCA The Fly

A small constellation in the south polar region of the sky. It was originally known as Apis, the Bee. Part of the dark Coalsack Nebula spills over into Musca from neighbouring Crux, the Southern Cross.

DOUBLE STARS

β (beta) Muscae appears to the naked eye as a blue-white star of mag. 3.0. Telescopes of 100 mm aperture and high magnification show that it actually consists of a close pair of 4th-mag. stars.

θ (theta) Muscae is a double of 6th- and 8th-mag. stars for small telescopes. The fainter component is a small, hot star of the type known as a Wolf–Rayet star (a rare class of stars with very hot surfaces which seem to be ejecting gas).

Musca, shown under its old name of Apis, the Bee, on the 1801 atlas of Johann Bode. Royal Greenwich Observatory

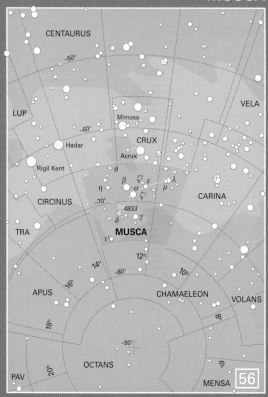

MUSCA

CENTAURUS

−50°

LUP

VELA

−60°

Mimosa

CRUX

Hadar

Acrux

Rigil Kent

θ

β ζ² ε λ

η α μ

ζ¹

CIRCINUS −70° γ CARINA

4833

δ

ι MUSCA

TRA

12h

14h −80°

16h 10h

APUS CHAMAELEON

VOLANS

18h

8h

−90°

OCTANS

20h 6h

PAV MENSA

56

NORMA The Set Square

A small constellation of the southern hemisphere of the sky, representing a draughtsman's set square; its brightest star is of only 4th mag. The constellation's boundaries have been changed since it was first formed, so that it no longer contains the stars that were originally labelled α (alpha) and β (beta).

Double stars

γ¹ γ² (gamma¹ gamma²) Normae are an unrelated pair of yellow giant and supergiant stars of 4th and 5th mags.

ε (epsilon) Normae is a 5th-mag. star with a 7th-mag. companion visible in small telescopes. Professional astronomers have found that the brighter star is a spectroscopic binary, making this a triple system.

ι¹ (iota¹) Normae is a 5th-mag. star with an 8th-mag. companion for small telescopes.

Open cluster

NGC 6087 is a large binocular cluster of about 40 stars, including the Cepheid variable S Normae which ranges from mag. 6.1 to 6.8 every 9.8 days.

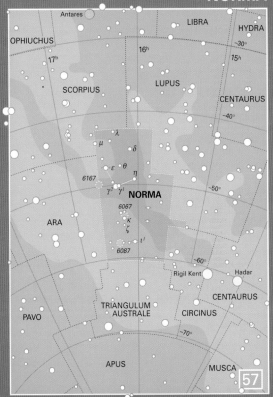

NORMA

Antares

OPHIUCHUS

LIBRA

HYDRA

~30°

17h

16h

15h

SCORPIUS

LUPUS

CENTAURUS

~40°

λ

μ

δ

θ

ε

η

6167

γ² γ¹

NORMA

~50°

6067

κ

ARA

ζ

ι¹

6087

~60°

Rigil Kent

Hadar

CENTAURUS

TRIANGULUM
AUSTRALE

CIRCINUS

PAVO

~70°

APUS

MUSCA

57

OCTANS The Octant

The constellation that contains the south celestial pole. Appropriately enough it represents an old instrument used for navigation, the octant, a forerunner of the sextant. Despite its privileged position, the constellation is not prominent; its brightest stars are of only 4th mag.

DOUBLE STAR
λ (lambda) Octantis is a double star for small telescopes, consisting of yellow and white components of 5th and 7th mags.

SOUTH POLE STAR
σ (sigma) Octantis is the nearest naked-eye star to the south celestial pole. It is a yellow-white giant star of mag. 5.4, 270 l.y. away. Currently σ Octantis lies about 1° from the celestial pole. It was closest to the celestial pole in the 19th century when it lay about 45′ away. The effect of precession is moving the celestial pole away from σ Octantis into a blank area of sky in the direction of Chamaeleon. The next reasonably bright southern pole star will be the wide 4th-mag. double δ (delta) Chamaeleontis in about 2000 years' time.
To find the exact position of the south celestial pole, note that it forms a near-equilateral triangle with χ (chi) and τ (tau) Octantis.

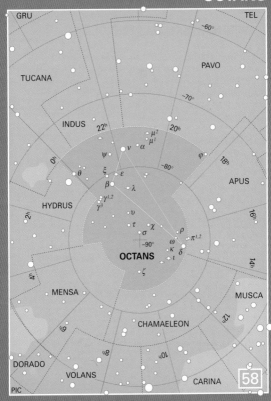

GRU
TEL
~-60°
TUCANA
PAVO
INDUS
~-70°
22h
μ²
20h
μ¹
ψ
ν
α
φ
18h
0h
ξ
ε
~-80°
θ
β
λ
APUS
γ³
γ¹,²
ν
HYDRUS
τ
χ
ρ
2h
σ
ω
π¹,²
16h
OCTANS
κ
δ
ι
14h
ζ
MENSA
MUSCA
12h
CHAMAELEON
DORADO
10h
VOLANS
CARINA
PIC

58

OPHIUCHUS The Serpent Holder

A large constellation of the equatorial region of the sky, representing a man encoiled by a serpent (the constellation Serpens). Ophiuchus is usually identified as Aesculapius, a mythical Greek healer. The Sun passes through Ophiuchus during December each year, but the constellation is not part of the zodiac.

DOUBLE AND MULTIPLE STARS

ρ (rho) Ophiuchi is a complex multiple star. Small telescopes with high magnification show it as a close pair of 5th- and 6th-mag. stars; in binoculars, two wide 7th-mag. companions can be seen, one on each side, forming a V-shaped grouping. ρ Ophiuchi is embedded in faint nebulosity which shows up only on long-exposure photographs.

τ (tau) Ophiuchi is a close pair of 5th- and 6th-mag. stars requiring 100 mm aperture and high magnification to be divided.

36 Ophiuchi is a neat pair of 5th-mag. orange stars for small telescopes.

70 Ophiuchi is a close but beautiful double star consisting of yellow and orange components of 4th and 6th mag., divisible in 75 mm aperture. At a distance of 16.6 l.y., 70 Ophiuchi is relatively near to us.

▶

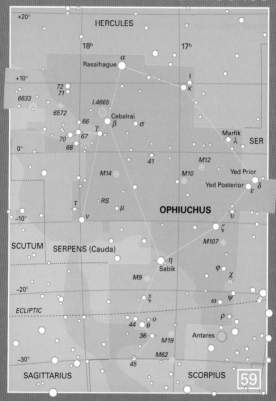

OPHIUCHUS

HERCULES

+20°

18h   17h

α
Rasalhague

+10°

ι
κ

72
71
6633

I.4665

6572
66
γ
Cebalrai
β
σ

70
67
68

Marfik
λ   SER

0°

41   M12

M14   M10   Yed Prior

τ   RS   M107
μ
Yed Posterior
δ
ε

OPHIUCHUS

ν
υ
ζ

SCUTUM

SERPENS (Cauda)   M107

-10°

η
Sabik   φ   χ
M9

ω   ψ
-20°

ECLIPTIC   ξ
o
ρ
44   θ
36   M19   Antares
45   M62

-30°

SAGITTARIUS   SCORPIUS   59

VARIABLE STAR

RS Ophiuchi is a recurrent nova that has flared from 11th to 5th mag. in 1898, 1933, 1958, 1967 and 1985, and can be expected to do so again.

NEARBY STAR

Barnard's Star, 5.9 l.y., away, is the next-closest star to the Sun after the triple system of α (alpha) Centauri. It is a red dwarf of mag. 9.5, within the range of small telescopes. The finder chart opposite will help you locate this elusive but fascinating object. Barnard's Star has the largest proper motion of any star; it moves across 1° of sky every 350 years.

OPEN CLUSTERS

IC 4665 is an easy binocular cluster consisting of a loose scattering of about 20 stars of mag. 7 and fainter covering a wider area of sky than the full Moon, visible in the same field as the 3rd-mag. orange giant β (beta) Ophiuchi.

NGC 6633 is a loose binocular cluster of 30 or so stars.

GLOBULAR CLUSTERS

M10 (NGC 6254) and M12 (NGC 6218), both of 7th mag., are the most prominent of several globular clusters visible in Ophiuchus. In binoculars or small telescopes they are seen as misty patches. Of the two, M10 appears the more condensed.

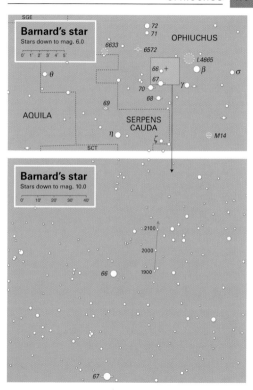

Barnard's star
Stars down to mag. 6.0

0' 1' 2' 3' 4' 5'

SGE

OPHIUCHUS

72
71

6633
6572

66 +
I.4665
β

θ
67
γ

70
σ

68

69

AQUILA

SERPENS
CAUDA

η
M14

ζ

SCT

Barnard's star
Stars down to mag. 10.0

0' 10' 20' 30' 40'

2100
2000
1900

66

67

ORION The Hunter

A magnificent constellation of the equatorial region of
the sky, representing a hunter or warrior with his shield
and club raised against the snorting charge of neighbouring Taurus the Bull. In Greek mythology, boastful
Orion was stung to death by a scorpion, and is now
placed in the sky so that he sets in the west as his slayer,
represented by the constellation Scorpius, rises in the
east.

BRIGHT STARS

α (alpha) Orionis (Betelgeuse) is a red supergiant star
and a semi-regular variable that fluctuates between
mags. 0.0 and 1.3 with a rough period of about 6 years.
At its maximum it is among the ten brightest stars in
the sky. It pulsates in diameter between about 300 and
400 times the size of the Sun. It lies 430 l.y. away.

β (beta) Orionis (Rigel, 'foot') is a blue-white supergiant of mag. 0.2, the seventh-brightest star in the sky
and the brightest in Orion. Rigel is 770 l.y. distant,
and has a 7th-mag. companion that is difficult for the
smallest telescopes to distinguish because of the over-
powering glare from Rigel itself.

γ (gamma) Orionis (Bellatrix, 'the warrior') is a blue
giant of mag. 1.6.

ε (epsilon) Orionis (Alnilam, 'string of pearls') is a
mag. 1.7 blue supergiant.

▶

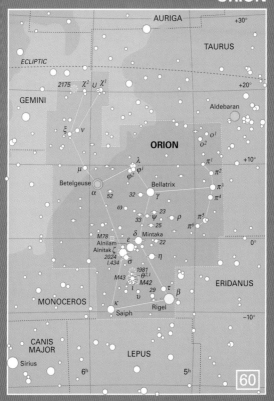

DOUBLE AND MULTIPLE STARS

δ (delta) Orionis (Mintaka, 'the belt') is a mag. 2.2 blue-white star with a wide 7th-mag. companion for binoculars and small telescopes.

ζ (zeta) Orionis (Alnitak, 'the girdle') consists of a tight pair of mags. 1.8 and 4.0 requiring at least 75 mm aperture and high magnification to be split. There is also a wider 10th-mag. companion.

θ¹ (theta¹) Orionis, also known as the Trapezium, is a multiple star at the heart of the Orion Nebula. This group of stars has been born from the gas of the Orion Nebula and their light makes it glow. Small telescopes show a rectangular arrangement of four stars ranging from 5th to 8th mag. There are also two other 11th-mag. stars in the group, visible with apertures of 100 mm.

θ² (theta²) Orionis lies near θ¹ (theta¹) Orionis, and is a binocular duo of 5th- and 6th-mag. stars.

ι (iota) Orionis is an unequal double star of 3rd and 7th mags. for small telescopes. Also visible in the same field is a wider double known as Struve 747, of 5th and 6th mags.

λ (lambda) Orionis is a tight pair of 4th- and 6th-mag. stars for small telescopes.

σ (sigma) Orionis is a stunning multiple star. Binoculars show that this 4th-mag. blue-white star has a 7th-mag. ▶

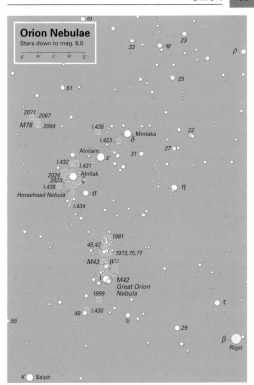

Orion Nebulae
Stars down to mag. 8.0

0'   30'   1'   90'   2'

ω

33   ψ   23

ρ

25

51

2071 2067

M78   2064

I.426   δ Mintaka

I.423

22

Alnilam   ε   31   27

I.432   I.431

2024   Alnitak   ζ

2023

I.435

Horsehead Nebula   σ   η

I.434

1981

45,42

1973,75,77

M43   θ²,¹

I   M42
Great Orion
Nebula

1999

49   I.430

υ   τ

55

29

β
Rigel

κ   Saiph

companion. Small telescopes reveal two closer companions of 7th and 9th mags., giving σ Orionis the appearance of a moon with planets. Also in the same telescopic field is the triple star Struve 761, consisting of a thin triangle of 8th- and 9th-mag. stars, completing a delightfully rich grouping.

OPEN CLUSTER

NGC 1981 is a scattered binocular cluster of about 20 stars including the double star Struve 750, a pair of 6th- and 8th-mag. stars for small telescopes.

NEBULAE

M42 (NGC 1976) and M43 (NGC 1982) together form one of the most celebrated objects in the entire heavens: the great Orion Nebula, a cloud of gas and dust fully 1° across in which stars are being born. The Nebula itself is visible to the naked eye as a misty haze making up the sword of Orion. It is prominent in binoculars and is breathtaking in telescopes under low powers which reveal indescribably complex twists and swirls of gas. Visually, the Orion Nebula seems to be in two parts, M42 and M43, but photographs show that these are both part of the same large cloud separated by a dark intrusion known as the Fish Mouth. To the human eye the Orion Nebula appears greenish, as do most nebulae, but photographs show that its true colour is reddish-orange. The reason for this colour difference is that the human eye has poor sensitivity to colours at such low light levels. At the heart of the

Orion Nebula is the multiple star θ^1 (theta[1]) Orionis (see page 182) which lights up the surrounding gas. The Orion Nebula lies 1500 l.y. away and has a diameter of 20 l.y. It is estimated to contain enough gas to make a cluster of thousands of stars.

NGC 1977 is a patch of nebulosity north of the Orion Nebula. It surrounds the 5th-mag. stars 42 and 45 Orionis. Although visible in small telescopes, NGC 1977 is often overlooked in favour of its more impressive neighbour M42.

The Horsehead Nebula is a strikingly shaped dark nebula looking like the black knight in a celestial chess game. It is caused by a cloud of dark dust overlying the tenuous nebulosity of IC 434 which stretches southwards from ζ (zeta) Orionis. However, IC 434 is visually highly elusive in even the largest amateur telescopes and the Horsehead itself is all but invisible, so observers must be content with views of this remarkable object provided by long-exposure photographs.

METEORS
The Orionid meteors reach their peak around October 22 each year, when as many as 25 meteors per hour may be seen coming from a point near the border with Gemini.

PAVO The Peacock

A modest constellation of the southern celestial hemi-sphere, representing the bird that was sacred to the goddess of the heavens, Juno.

BRIGHT STAR

α (alpha) Pavonis (Peacock) is a blue-white star of mag. 1.9.

VARIABLE STAR

κ (kappa) Pavonis is one of the brightest Cepheid variables, a yellow-white supergiant that varies between mags. 3.9 and 4.8 every 9.1 days.

DOUBLE STAR

ξ (xi) Pavonis is an unequal double star consisting of a 4th-mag. red giant with a 9th-mag. companion, difficult to see in small telescopes because it becomes lost in the primary's glare.

GLOBULAR CLUSTER

NGC 6752 is a large 5th-mag. globular cluster for binoculars and small telescopes. Within its outer regions lies a tight 8th-mag. double star but this is a foreground object.

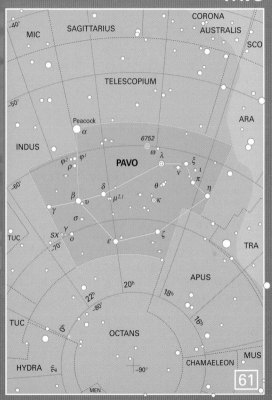

61

PEGASUS The Winged Horse

A large constellation in the northern half of the sky,
representing the winged horse of Greek mythology.
A feature of the constellation is the Great Square of
Pegasus whose corners are marked out by four stars;
one of these stars, originally known as δ (delta) Pegasi,
is now assigned to neighbouring Andromeda.

VARIABLE STAR

β (beta) Pegasi (Scheat, 'shin') is a red giant irregular
variable that fluctuates between 2nd and 3rd mag. with
no set period.

DOUBLE STARS

ε (epsilon) Pegasi (Enif, 'nose') is a double star with
components of widely unequal brightness. The main
star is a 2nd-mag. orange supergiant; its companion,
visible in small telescopes or binoculars, is of 8th mag.

1 Pegasi is a 4th-mag. yellow giant with an 8th-mag.
companion for small telescopes.

GLOBULAR CLUSTER

M15 (NGC 7078) is one of the finest globular clusters
in the northern sky. It is a 6th-mag. object visible in
binoculars and small telescopes, which show it as a
glorious misty patch in an attractive field. Apertures of
150 mm are needed to resolve individual stars.

PEGASUS

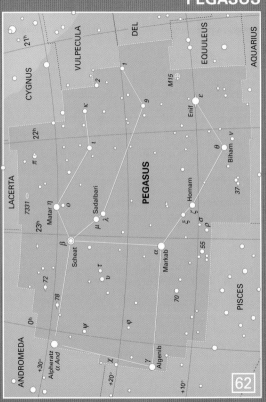

21ʰ

VULPECULA

DEL

EQUULEUS

AQUARIUS

CYGNUS

1

2

M15

ε

Enif

9

κ

22ʰ

ι

θ

ν

Biham

π

LACERTA

7331

Matter η

ο

μ

λ

Sadalbari

PEGASUS

37

ξ

Homam

ζ

σ

ρ

23ʰ

β

Scheat

α

Markab

55

τ

ν

70

PISCES

+30°

0ʰ

ψ

φ

ANDROMEDA

Alpheratz
α And

χ

γ

Algenib

+20°

+10°

62

PERSEUS

A constellation of the northern sky representing the hero of Greek mythology who saved Andromeda from being devoured by the sea monster Cetus. In the sky, Perseus is depicted as holding the severed head of Medusa the Gorgon, marked by the winking star β (beta) Persei.

BRIGHT STAR

α (alpha) Persei is a yellow-white supergiant of mag. 1.8. It lies within a widely scattered cluster, Melotte 20, that forms an attractive star field in binoculars.

VARIABLE STARS

β (beta) Persei (Algol, 'the demon') is the prototype of the eclipsing binary variables. These consists of two stars in close orbit that periodically pass in front of each other as seen from Earth, causing a drop in observed light. Algol itself varies from mag. 2.1 to 3.4 every 2 days 21 hours. For a chart, see page 193.

ρ (rho) Persei is a red giant semi-regular variable that fluctuates between 3rd and 4th mag. approximately every 7 weeks.

DOUBLE STARS

ε (epsilon) Persei is a 3rd-mag. blue-white star with an 8th-mag. companion, difficult in the smallest telescopes because of the brightness contrast.

▶

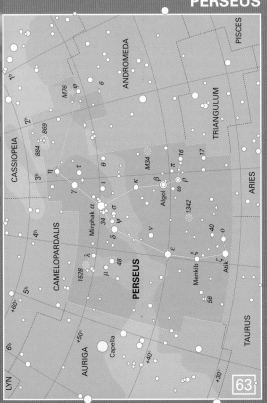

PISCES

ANDROMEDA

CASSIOPEIA

TRIANGULUM

M76

φ

6

η τ θ

M34 π

16

17

ARIES

884 869

γ κ β

ι Algol

ω ρ

34 σ

Mirphak α

δ ψ

1342

CAMELOPARDALIS

ν

40 ο

1528

λ

ε

μ 48

ξ ζ

PERSEUS

Menkib 56 Atik

TAURUS

AURIGA

Capella

LYN

63

ζ (zeta) Persei is a 3rd-mag. blue supergiant with a 9th-mag. companion for small telescopes.

η (eta) Persei is an attractive pair for small telescopes, consisting of orange and blue stars of 4th and 9th mags. in a star-sprinkled field.

OPEN CLUSTERS

NGC 869 and NGC 884, also known as h and χ (chi) Persei, are the famous Double Cluster, a related pair of star clusters that forms one of the richest sights in the sky for small instruments. To the naked eye the two clusters appear like a brightening in the Milky Way. Binoculars show that they both consist of a scattering of bright stars, each cluster covering more than ½° of sky, with NGC 869 being the brighter and richer of the pair. A few red stars can be seen in NGC 884 with small telescopes, but none in NGC 869. Both clusters lie about 7400 l.y. away in a spiral arm of our Galaxy.

M34 (NGC 1039) is a large 5th-mag. cluster, easily resolved into individual stars by small telescopes.

METEORS

The Perseid meteors are the most glorious meteor shower of the year, producing a maximum of as many as 80 meteors per hour on August 12 or 13 each year. Perseid meteors are bright, many of them exploding and leaving trains.

Right: Comparison chart and light curve for Algol

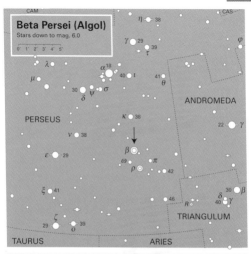

Beta Persei (Algol)
Stars down to mag. 6.0

0° 1° 2° 3° 4° 5°

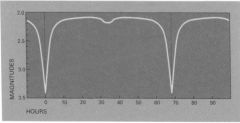

PHOENIX The Phoenix

A constellation of the southern celestial hemisphere,
representing the mythological bird that was reborn
from its own ashes.

VARIABLE STAR

ζ (zeta) Phoenicis is a complex variable and double star
(see below), the brightest component of which is an
eclipsing binary that varies between mags. 3.9 and 4.4
every 40 hours.

DOUBLE AND MULTIPLE STARS

β (beta) Phoenicis is a tight binary of 4th-mag. yellow
stars, currently too close to be resolved by amateur telescopes but within range of 100 mm apertures when at
their widest.

ζ (zeta) Phoenicis is a double star as well as a variable
star. Small telescopes show it as a 4th-mag. blue-white
star with an 8th-mag. companion; the brighter of the
two stars is an eclipsing binary (see above).

PHOENIX

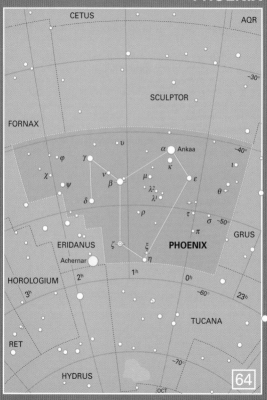

CETUS

AQR

SCULPTOR

FORNAX

~-30°

υ

φ

γ

α Ankaa

~-40°

ι

χ

ν

κ

μ

ε

θ

ψ

β

λ²

δ

λ

ρ

τ

σ

~-50°

π

ζ

ξ

η

PHOENIX

ERIDANUS

GRUS

Achernar

HOROLOGIUM

2h

1h

0h

~-60°

23h

3h

RET

TUCANA

HYDRUS

~-70°

OCT

64

PICTOR The Painter's Easel

A faint and insignificant constellation of southern skies. Its original title was Equuleus Pictoris, which has since been shortened to simply Pictor. Pictor contains little to interest the casual observer.

NEARBY STAR
Kapteyn's Star is a 9th-mag. red dwarf, lying 12.8 l.y. away. It is noted for having the second-largest proper motion of any star. (The record for the largest proper motion of all is held by Barnard's Star in Ophiuchus – see page 178). Kapteyn's Star takes 415 years to move across 1° of sky.

PLANETARY SYSTEM?
β (beta) Pictoris is a 4th-mag. star 63 l.y. away which is notable for being encircled by a disk of dust, detectable by professional instruments, from which planets are thought to be forming.

PICTOR

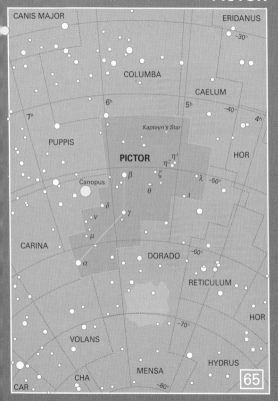

CANIS MAJOR

ERIDANUS

~-30°

COLUMBA

CAELUM

6h

5h

~-40°

7h

4h

Kapteyn's Star

PUPPIS

η¹

PICTOR

η²

HOR

Canopus

β

ζ

λ

~-50°

θ

ι

δ

γ

ν

μ

CARINA

α

DORADO

~-60°

RETICULUM

HOR

VOLANS

~-70°

HYDRUS

CAR

CHA

MENSA

~-80°

65

PISCES The Fishes

A constellation of the zodiac through which the Sun passes from mid-March to late April. The Sun crosses the celestial equator from south to north while in Pisces; this point is known as the vernal equinox and is reached around March 20 each year. Pisces represents a pair of fishes tied together by their tails. It is a surprisingly faint constellation, its brightest stars being of only 4th mag.

VARIABLE STAR
TX Piscium (19 Piscium) is a red giant irregular variable that fluctuates around 5th mag. It is notable for its deep red colour.

DOUBLE STARS
α (alpha) Piscium (Alrescha, 'the cord') is a close pair of 4th- and 5th-mag. white stars requiring 100 mm aperture to see separately. The stars are currently closing together and hence becoming progressively more difficult to divide.

ζ (zeta) Piscium is an easy pair of 5th- and 6th-mag. stars for small telescopes.

ρ (rho) Piscium and 94 Piscium form an easy 5th-mag. binocular pair of white and orange stars.

$ψ^1$ (psi^1) Piscium is a wide pair of 5th- and 6th-mag. blue-white stars for small telescopes. ▶

PISCES

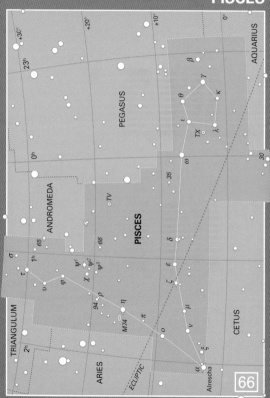

AQUARIUS

PEGASUS

ANDROMEDA

PISCES

TRIANGULUM

ARIES

CETUS

Alrescha

ECLIPTIC

M74

TV

TX

β

γ

κ

θ

ι

λ

ω

δ

ε

ζ

μ

ν

ο

ξ

α

π

η

ρ

χ

ψ¹

ψ²

ψ³

φ

υ

τ

σ

94

65

66

+30°

+20°

+10°

0°

23ʰ

0ʰ

1ʰ

2ʰ

30

35

66

GALAXY

M74 (NGC 628) is a 9th-mag. spiral galaxy presented face-on, appearing impressive on long-exposure photographs but too faint to be seen well in small amateur telescopes.

PISCIS AUSTRINUS The Southern Fish

A constellation of the southern sky, unremarkable apart from its brightest star, Fomalhaut. In mythology it was supposedly the parent of the zodiacal fishes represented by Pisces. The constellation is sometimes also called Piscis Australis.

BRIGHT STAR

α (alpha) Piscis Austrini (Fomalhaut, from the Arabic meaning 'fish's mouth') is a white star of mag. 1.2, lying 25 l.y. away.

DOUBLE STARS

β (beta) Piscis Austrini is a 4th-mag. white star with an 8th-mag. companion that can be picked up in small telescopes.

γ (gamma) Piscis Austrini is a pair of 4th- and 8th-mag. stars, difficult to separate in small telescopes because of the contrast in brightness.

PISCIS AUSTRINUS

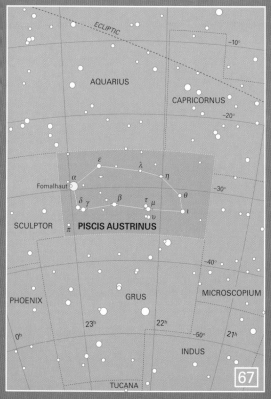

ECLIPTIC

−10°

AQUARIUS

CAPRICORNUS

−20°

ε

λ

α

η

Fomalhaut

−30°

δ γ

β

τ μ

θ

υ

ι

SCULPTOR

π

PISCIS AUSTRINUS

−40°

PHOENIX

GRUS

MICROSCOPIUM

23h

22h

0h

−50°

21h

INDUS

TUCANA

67

PUPPIS The Stern

A major southern constellation, formerly part of the ancient Greek constellation Argo Navis, the Ship of the Argonauts. It contains rich Milky Way star fields.

VARIABLE STARS

L^2 Puppis is a red giant semi-regular variable that fluctuates between 3rd and 6th mag. about every 140 days. It is unrelated to L^1 Puppis.

V Puppis is an eclipsing binary that varies from mag. 4.4 to mag. 4.9 every 35 hours.

DOUBLE STARS

ξ (xi) Puppis is a 3rd-mag. yellow supergiant with an unrelated 5th-mag. yellow binocular companion.

k Puppis is a neat pair of 4th- and 5th-mag. blue-white stars for small telescopes.

OPEN CLUSTERS

M46 (NGC 2437), M47 (NGC 2422) are prominent binocular clusters, seemingly adjacent but at different distances, jointly visible to the naked eye as a brighter knot in the Milky Way.

NGC 2451 is a scattered group of stars surrounding the 4th-mag. orange giant c Puppis.

NGC 2477 is a large and rich 6th-mag. cluster, appearing hazy in binoculars.

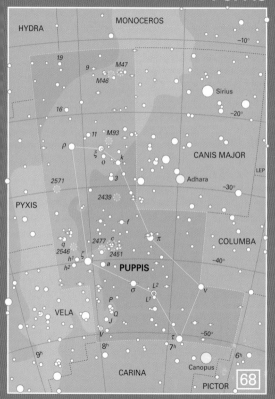

HYDRA

MONOCEROS

Sirius

−10°

19

9   M47

M46

16

−20°

11   M93

CANIS MAJOR

ρ

ζ

O   k

3

LEP

2571

Adhara

−30°

PYXIS

2439

f

2477   c

π

COLUMBA

q

2546

ζ

2451

h¹

a   **PUPPIS**

h²

σ

L²

−40°

ν

L¹

P

VELA

Q

J

−50°

V

τ

9ʰ

8ʰ

7ʰ

6ʰ

CARINA

Canopus

PICTOR

68

PYXIS The Compass

The smallest of the four parts into which the large
southern constellation of Argo Navis, the Ship of the
Argonauts, was divided; the other parts are Carina,
Puppis and Vela. The brightest stars of Pyxis are of
only 4th mag. and the constellation contains little to
interest the casual observer.

RECURRENT NOVA

T Pyxidis is a recurrent nova that has been seen to
erupt five times (in 1890, 1902, 1920, 1944 and 1966), a
record that it shares with RS Ophiuchi. At its brightest
it can reach 6th mag. but normally it slumbers at 14th
mag. Further outbursts may be expected.

RETICULUM The Net

A small and insignificant constellation of the southern
hemisphere of the sky, representing an instrument used
by astronomers for measuring star positions. Its brightest star is of 3rd mag.

DOUBLE STAR

ζ (zeta) Reticuli is a naked-eye or binocular pair of
near-identical 5th-mag. yellow stars similar to our own
Sun. They lie 39 l.y. away.

PYXIS

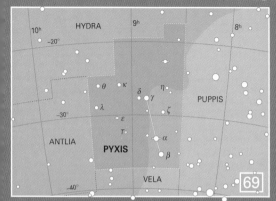

RETICULUM

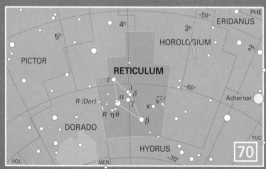

SAGITTA The Arrow

A constellation with a distinctive arrow shape, lying in the northern celestial hemisphere. It is the third-smallest constellation, and its brightest stars are of only 4th mag. One legend says that it represents an arrow shot by neighbouring Hercules. Sagitta lies in a rich part of the Milky Way.

VARIABLE STARS
S Sagittae is a Cepheid variable that fluctuates between mags. 5.2 and 6.0 every 8.4 days.

VZ Sagittae is a red giant that varies irregularly from about mag. 5.3 to 5.6.

DOUBLE STAR
ζ (zeta) Sagittae is a pair of 5th- and 9th-mag. stars for small telescopes.

RECURRENT NOVA
WZ Sagittae is a nova that has flared up from 15th to 7th or 8th mag. in 1913, 1946, 1978 and 2001. Further outbursts may occur at any time.

GLOBULAR CLUSTER
M71 (NGC 6838) is an 8th-mag. globular cluster for binoculars and small telescopes which show it as a rounded misty patch, somewhat elongated.

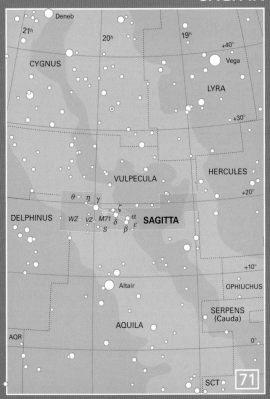

SAGITTA

21h

Deneb

CYGNUS

20h

19h

+40°

Vega

LYRA

+30°

VULPECULA

HERCULES

+20°

θ η γ
ζ α
δ ε

DELPHINUS

WZ VZ M71

SAGITTA

S
β

+10°

Altair

OPHIUCHUS

SERPENS
(Cauda)

AQUILA

AQR

0°

SCT

71

SAGITTARIUS The Archer

The Sun lies in this zodiacal constellation from mid-December to mid-January. Sagittarius represents a centaur aiming a bow and arrow at the heart of the neighbouring scorpion, Scorpius. The constellation's main stars are often visualized as forming the shape of a teapot. Sagittarius contains rich Milky Way star fields towards the centre of our Galaxy, plus numerous clusters and nebulae.

VARIABLE STARS
W Sagittarii is a Cepheid variable that fluctuates between mags. 4.3 and 5.1 every 7.6 days.

X Sagittarii is another Cepheid that varies from mag. 4.2 to 4.9 every 7 days.

MULTIPLE STAR
β (beta) Sagittarii is a naked-eye double, consisting of two unrelated 4th-mag. stars. β^1 (beta^1) Sagittarii, the more northerly of the pair, has a 7th-mag. companion visible in small telescopes.

OPEN CLUSTERS
M23 (NGC 6494) is a large binocular cluster the apparent size of the full Moon. Small telescopes resolve some of its 150 or so stars of mag. 9 and fainter. ▶

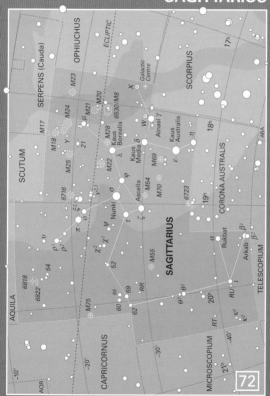

M24 is not really a true star cluster; rather, it is the richest part of the Milky Way in Sagittarius, an elongated star cloud 2° long and 1° wide, visible to the naked eye. Binoculars and telescopes show it as a field of stardust.

M25 (IC 4725) is a large, scattered binocular cluster of about 30 stars, of 7th mag. and fainter. Near the cluster's centre is the yellow supergiant U Sagittarii, a Cepheid variable that ranges from mag. 6.3 to mag. 7.1 every 6.7 days.

NGC 6530 is a binocular cluster of about 25 stars of mag. 7 and fainter lying in the Lagoon Nebula, M8 (see below). The stars of the cluster have apparently formed recently from the surrounding gas.

GLOBULAR CLUSTER
M22 (NGC 6656) is a 5th-mag. cluster visible to the naked eye and prominent in binoculars. It is rated as one of the finest globular clusters in the sky. Small telescopes reveal its noticeably elliptical outline, but apertures of at least 75 mm are needed to resolve individual stars within it. M22 is one of the closest globulars to us, some 10,000 l.y. away.

NEBULAE
M8 (NGC 6523), the Lagoon Nebula, is the best of the diffuse nebulae in Sagittarius and one of the finest in the entire sky. It is visible to the naked eye as a 5th-mag. hazy patch of similar size to the Orion Nebula.

Binoculars and small telescopes show the cluster NGC 6530 (see facing page) that lies within it. The Nebula is divided by a dark rift, the so-called Lagoon that gives the object its name, but apertures of 75 mm and above are needed to show this well. On the other side of the rift from the cluster NGC 6530 are two prominent stars, the brighter of which is 6th-mag. blue-white supergiant 9 Sagittarii. Farther to the side is the 5th-mag. yellow giant 7 Sagittarii, but this is a foreground object and not associated with the nebula. The full complexity of the Lagoon Nebula, and its pinkish-red colour caused by hydrogen gas, are brought out only on photographs (see page 20).

M17 (NGC 6618) is known variously as the Omega Nebula, the Swan Nebula and the Horseshoe Nebula on account of its looped shape, although this is defined only in larger telescopes. In binoculars and small telescopes it appears as an elongated smudge like the tail of a comet, with a sprinkling of faint stars along one side. This unnamed mini-cluster is apparently associated with the nebula.

M20 (NGC 6514), the Trifid Nebula, is a famous cloud of gas that takes its name from the lanes of dust that tri-sect it as seen on photographs. But small telescopes show only the 7th-mag. double star at its heart. Larger apertures are needed to see the nebulosity and dark dividing lanes. Note also the loose cluster M21 (NGC 6531) in the same field of view.

SCORPIUS The Scorpion

A magnificent constellation of the zodiac, through which the Sun passes during the last week of November. It represents the scorpion that killed Orion with its sting.

BRIGHT STAR

α (alpha) Scorpii (Antares, from the Greek meaning 'like Mars', in reference to its strong red colour) is a red supergiant perhaps 500 times the diameter of the Sun. It is also a semi-regular variable that fluctuates from about mag. 0.9 to mag. 1.2 over a period of 5 years or so. Antares has a 6th-mag. blue-green companion which requires at least 75 mm aperture and steady air to be picked out from the glare of its primary.

DOUBLE AND MULTIPLE STARS

β (beta) Scorpii (Graffias, 'claws') is an impressive pair of 3rd- and 5th-mag. stars for small telescopes.

$ζ^1 ζ^2$ (zeta^1 zeta^2) Scorpii is a naked-eye pair of unrelated stars. $ζ^1$ is a 5th-mag. blue-white supergiant that is probably an outlying member of the cluster NGC 6231 (see page 214) while $ζ^2$ is a 4th-mag. orange giant much closer to us.

$μ^1 μ^2$ (mu^1 mu^2) Scorpii is a naked-eye pair, consisting of two blue-white stars of 3rd and 4th mag. $μ^1$ is an eclipsing binary with a range of only 0.3 mag. and a period of 1.4 days.

▶

SCORPIUS

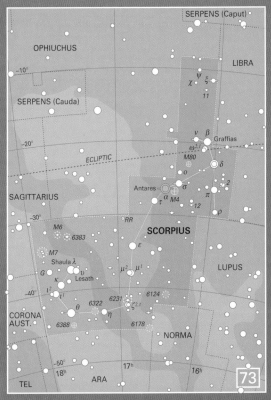

SERPENS (Caput)

OPHIUCHUS

LIBRA

ψ
χ   ξ
11

SERPENS (Cauda)

−10°

ν   β
ω²,¹   Graffias
M80
δ

ECLIPTIC

−20°

σ   τ 2
π

SAGITTARIUS

Antares   ○
α   M4   12
τ

−30°   ρ

RR

M6

SCORPIUS

6383

ε

M7

LUPUS

Shaula λ   μ²   μ¹
G
κ   υ
Lesath
−40°   ι²
ι¹   6231   6124
θ   6322   ζ²,¹
CORONA   η   6178
AUST.   6388

NORMA

−50°

18ʰ   17ʰ   16ʰ

TEL   ARA

73

ν (nu) Scorpii appears as a wide pair of 4th- and 6th-mag. stars in small telescopes. Apertures of at least 75 mm show that the fainter star is itself double.

ξ (xi) Scorpii is an excellent quadruple star for small telescopes, which show it as a 4th-mag. star with a 7th-mag. orange companion. Also visible in the same field is a wider duo of 7th- and 8th-mag. stars, called Struve 1999, which are gravitationally connected to the first pair.

$ω^1$ $ω^2$ (omega[1] omega[2]) Scorpii is a naked-eye pair of unrelated 4th-mag. stars.

OPEN CLUSTERS

M6 (NGC 6405) is a 4th-mag. object, sometimes called the Butterfly Cluster, consisting of about 80 stars of mag. 7 and fainter seemingly arranged in radiating chains. Its brightest star is an orange giant variable, BM Scorpii, which ranges between 5th and 7th mag.

M7 (NGC 6475) is a large and bright cluster visible to the naked eye covering 1° of sky. It is an outstanding binocular object, consisting of about 80 members of 6th mag. and fainter, arranged in a cruciform shape.

NGC 6231 is a naked-eye cluster of about 120 stars of 6th to 8th mag., like a mini-Pleiades, on a background of many fainter stars. The 5th-mag. star $ζ^1$ (zeta[1]) Scorpii is thought to be an outlying member of the cluster. This is a glorious area for sweeping with binoculars; note a line of stars leading northwards from NGC 6231 to the very large and scattered cluster H12. The two clusters lie about 6000 l.y. away, and the stars joining them outline a spiral arm of our Galaxy.

GLOBULAR CLUSTER

M4 (NGC 6121) is a large 6th-mag. globular cluster, among the closest to us, under 7000 l.y. away. It is visible in binoculars in dark skies.

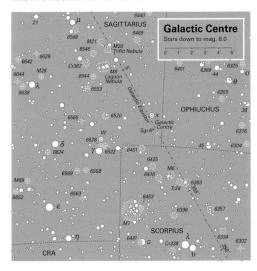

SCULPTOR The Sculptor

A faint constellation of the southern celestial hemisphere whose brightest stars are of only 4th mag. Sculptor contains the south galactic pole, i.e. the point 90° south of the plane of the Galaxy (the equivalent northern point lies in Coma Berenices).

VARIABLE STAR

R Sculptoris is a red giant semi-regular variable noted for its deep red colour. It varies between 6th and 8th mags. approximately every 370 days.

DOUBLE STARS

ε (epsilon) Sculptoris is a pair of 5th- and 9th-mag. stars for small telescopes.

κ¹ (kappa¹) Sculptoris is a tight pair of 6th-mag. stars requiring apertures of at least 75 mm and very high magnification.

GALAXIES

NGC 55 is an 8th-mag. spiral galaxy presented nearly edge-on to us. It can be picked up in small telescopes as an elongated smudge of uneven brightness.

NGC 253, the brightest of the galaxies in Sculptor, is a 7th-mag. spiral visible in binoculars. It is tilted at an angle to us, so that it appears cigar-shaped in small instruments.

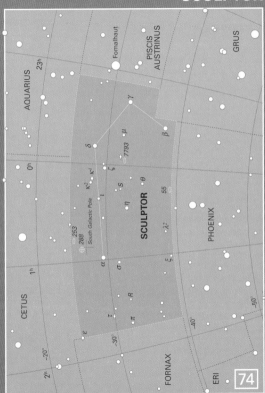

74

SCUTUM The Shield

This constellation lies just south of the celestial equator and is impressive despite its small size (the fifth-smallest constellation). It contains rich Milky Way star fields and is an attractive area for sweeping with binoculars.

VARIABLE STARS

δ (delta) Scuti is the prototype of a rare class of variable stars that undergo small pulsations in size every few hours, leading to minor changes in their brightness. δ Scuti itself varies every 4 hours 40 minutes from mag. 4.6 to 4.8, an amount that is barely detectable to the naked eye.

R Scuti is a somewhat unusual yellow supergiant variable star of the RV Tauri variety. It fluctuates semi-regularly between 4th and 9th mag.

OPEN CLUSTER

M11 (NGC 6705), the Wild Duck Cluster, is a beautiful and rich group of about 200 faint stars in a dense area of the Milky Way. It gets its name because its brightest stars are arranged in a fan-shape, like a flight of migrating ducks, with a brighter 8th-mag. orange star at the apex. It is visible in all apertures, including binoculars, but larger instruments will resolve the individual stars more clearly.

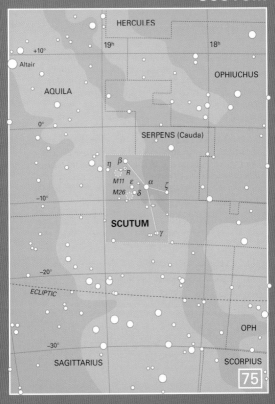

HERCULES

19ʰ

18ʰ

+10°

Altair

OPHIUCHUS

AQUILA

0°

SERPENS (Cauda)

η β
R
M11 ε α ζ
M26 δ

−10°

SCUTUM

γ

−20°

ECLIPTIC

OPH

−30°

SAGITTARIUS

SCORPIUS

75

SERPENS The Serpent

This is a unique constellation, for it is split into two separate halves (see map on pp. 222–223). Serpens represents a snake coiled around the serpent holder, Ophiuchus. One side of Ophiuchus lies Serpens Caput, the serpent's head, which is the larger and more prominent half; on the other side of Ophiuchus lies Serpens Cauda, the serpent's tail. Despite being split in this way, the two halves of Serpens are counted as one constellation.

DOUBLE STARS

β (beta) Serpentis consists of a pair of 4th- and 10th-mag. stars divisible in small telescopes. Binoculars show an unrelated 7th-mag. star nearby.

δ (delta) Serpentis is a close pair of 4th- and 5th-mag. stars for small telescopes with high magnification.

θ (theta) Serpentis is an easy pair of 5th-mag. white stars for small telescopes.

ν (nu) Serpentis is a 4th-mag. star with a wide 8th-mag. companion for small telescopes and binoculars.

OPEN CLUSTERS

τ^1 (tau[1]) Serpentis, a 5th mag. red giant, is the brightest of a scattered group of stars, best seen in binoculars, that extends towards κ (kappa) Serpentis.

The Eagle Nebula is a faint cloud of gas surrounding the star cluster M16 in Serpens. Bill Schoening/AURA/NOAO/NSF

M16 (NGC 6611) is a cluster of over 50 stars of mag. 8 and fainter, visible in binoculars and small telescopes. It lies in a rich Milky Way region near the borders with Scutum and Sagittarius. Under good conditions a hint of nebulosity may be detected around the cluster. This is brought out well on long-exposure photographs (see above) which show M16 to be buried in a glorious cloud of glowing gas known as the Eagle Nebula. A feature of this nebula is the ribbon of darker dust that intrudes into it.

GLOBULAR CLUSTER

M5 (NGC 5904) is a 6th-mag. globular cluster rated as second-finest in the northern skies to M13 in Hercules. M5 is within range of binoculars and small telescopes, which show it as a hazy star like the head of a comet. Larger telescopes reveal chains of stars radiating outwards from its bright and dense centre. In the same field of view is the 5th-mag. star 5 Serpentis.

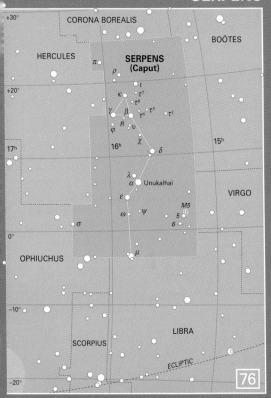

SERPENS
(Caput)

CORONA BOREALIS

BOÖTES

HERCULES

π

ρ

ι

κ

τ⁷

τ⁸

γ

β

τ⁶ τ⁵

τ¹

R

φ

υ

χ

δ

λ

α Unukalhai

ε

ω ψ

M5

5

σ

ε

VIRGO

μ

OPHIUCHUS

LIBRA

SCORPIUS

ECLIPTIC

+30°

+20°

+10°

0°

−10°

−20°

17ʰ

16ʰ

15ʰ

76

SEXTANS The Sextant

A barren constellation in the equatorial region of the sky, almost indistinguishable on account of its faintness. It represents the sextant, an instrument used by astronomers for measuring the positions of stars before the days of the telescope. The brightest star in Sextans is of only mag. 4.5.

DOUBLE STAR
17–18 Sextantis is a binocular pair of unrelated 6th-mag. stars, orange and blue-white.

GALAXY
NGC 3115 is a small 9th-mag. elliptical galaxy known as the Spindle Galaxy because of its shape. Moderate-sized telescopes are needed to show it well.

Sextans, shown under its original title Sextans Uraniae, on the 1801 atlas of Johann Bode. Royal Greenwich Observatory

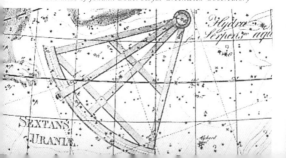

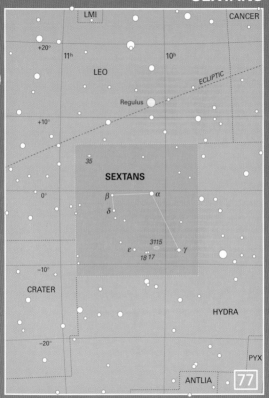

CANCER

LMI

+20°

11ʰ

LEO

10ʰ

ECLIPTIC

Regulus

+10°

35

SEXTANS

β

α

δ

0°

3115

ε

18 17

γ

-10°

CRATER

HYDRA

-20°

PYX

ANTLIA

TAURUS The Bull

A constellation of the zodiac, through which the Sun passes from mid-May to late June. It represents the head and shoulders of a bull, depicted as charging at neighbouring Orion.

BRIGHT STAR

α (alpha) Tauri (Aldebaran, from the Arabic meaning 'the follower', i.e. of the Pleiades star cluster) is an orange-coloured giant of mag. 0.9, the 14th-brightest star in the sky. It represents the glinting red eye of the Bull. Aldebaran lies 65 l.y. away, so that although it appears to be part of the Hyades star cluster (see p. 228) it is actually much closer to us.

VARIABLE STAR

λ (lambda) Tauri is an eclipsing binary that varies from mag. 3.4 to mag. 3.9 every 4 days.

DOUBLE STARS

$\theta^1 \theta^2$ (theta^1 theta^2) Tauri is a wide double star in the Hyades (see p. 228), divisible by binoculars or naked eye, consisting of white and yellow giants of 3rd and 4th mag.

$\kappa^1 \kappa^2$ (kappa^1 kappa^2) Tauri is a wide naked-eye or binocular double of 4th and 5th mags.

$\sigma^1 \sigma^2$ (sigma^1 sigma^2) Tauri are a wide binocular pair of 5th-mag. white stars.

▶

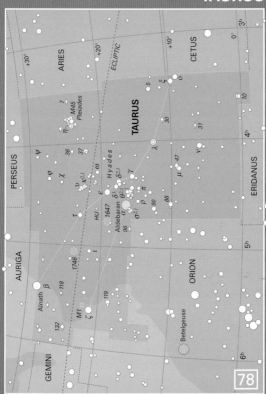

φ (phi) Tauri is a 5th-mag. orange giant with an unrelated 8th-mag. companion visible in small telescopes.

χ (chi) Tauri is an attractive pair of 5th- and 8th-mag. blue and gold stars for small telescopes.

OPEN CLUSTERS

The Hyades is a very large and scattered cluster of stars, arranged in a V-shape, that makes up the face of the Bull. The cluster contains about 200 stars, the brightest dozen or so of which are visible to the naked eye. Because of its considerable size, covering 5° of sky, the Hyades is best studied with binoculars. The cluster lies 150 l.y. away. Note that the bright star Aldebaran is not a member of the cluster but is actually a foreground object at less than half the distance.

M45, the Pleiades, is the finest star cluster in the sky. To the naked eye it appears as a slightly hazy group of about six stars, but binoculars and small telescopes bring many more stars into view, covering more than 1° of sky. In all, about 100 stars belong to the cluster, which lies 380 l.y. away, over twice as far as the Hyades. The brightest member of the Pleiades cluster is 3rd-mag. η (eta) Tauri (Alcyone). Under very clear conditions some faint nebulosity may be seen in binoculars and small telescopes around 23 Tauri (Merope). Long-exposure photographs show that the whole cluster is enveloped in faint nebulosity. ▶

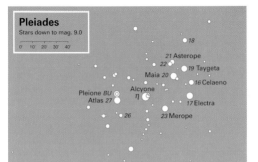

Pleiades
Stars down to mag. 9.0

0' 10' 20' 30' 40'

18
21 Asterope
22
19 Taygeta
Maia 20
16 Celaeno
Pleione *BU*
Alcyone
Atlas 27
η
17 Electra
26
23 Merope

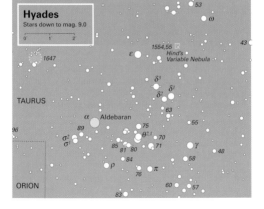

Hyades
Stars down to mag. 9.0

0' 1' 2'

53
ω
43
1554,55
Hind's
Variable Nebula
ε
1647
δ³
δ¹
δ²
TAURUS
63
α Aldebaran
55
89
75
96
σ²
θ²,¹
70
γ
48
σ¹
85 80
71
58
81
84
ρ
76
π
ORION
60 57
83

NEBULA
M1 (NGC 1952) is the Crab Nebula, the remains of a star that was seen to explode as a supernova in AD 1054, and one of the most famous objects in the heavens. It is a challenging object for small telescopes, appearing as an elliptical 8th-mag. hazy patch, but can be glimpsed in binoculars under good conditions. Its popular name comes from its supposed resemblance to the claws of a crab. Although not as spectacular in small instruments as it appears on photographs, it is worth seeking out as the brightest example of a supernova remnant.

METEORS
The Taurid meteor shower reaches a broad peak in the first half of November, when up to 10 meteors an hour can be seen coming from the region of the Hyades and Pleiades clusters. Taurid meteors are slow-moving and often bright.

TELESCOPIUM The Telescope

An unremarkable constellation of the southern sky, a poor tribute to the astronomer's most fundamental instrument, the telescope.

DOUBLE STAR
δ^1 δ^2 (delta^1 delta^2) Telescopii are a binocular pair of unrelated 5th-mag. blue-white stars.

TELESCOPIUM

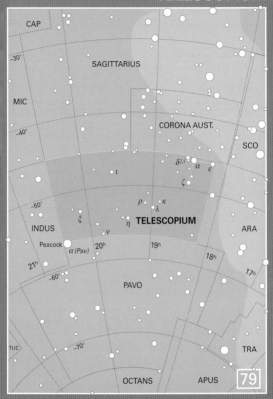

CAP

SAGITTARIUS

−30°

MIC

CORONA AUST.

−40°

SCO

δ²·¹ α ε

ζ

ρ κ

λ

ξ

η **TELESCOPIUM**

ν

INDUS

ARA

Peacock

α (Pav)

20ʰ

19ʰ

18ʰ

21ʰ

17ʰ

−60°

PAVO

TUC

−70°

TRA

OCTANS

APUS

79

TRIANGULUM The Triangle

A small constellation of distinctive triangular shape in the northern sky.

DOUBLE STAR
6 Trianguli is an attractive pair of 5th- and 7th-mag. golden and blue stars, divisible at high magnification in small telescopes.

GALAXY
M33 (NGC 598) is the third-largest member of our Local Group of galaxies, after the Andromeda Galaxy and our own Milky Way. It is large, covering more sky than the full Moon, but faint. M33 is best seen in binoculars or wide-field telescopes with low magnification, because of its low contrast against the sky background. A clear, dark night is vital to find M33.

TRIANGULUM AUSTRALE
The Southern Triangle

A small but easily identified southern constellation. Its three main stars are of 2nd and 3rd mags.

OPEN CLUSTER
NGC 6025 is a binocular group of about 60 stars of mag. 7 and fainter.

TRIANGULUM

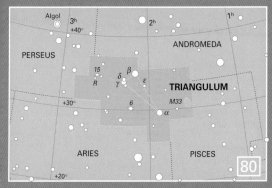

TRIANGULUM AUSTRALE

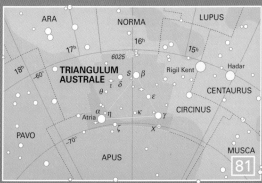

TUCANA The Toucan

A constellation of the southern sky, containing few bright stars but marked by the presence of the Small Magellanic Cloud and the globular cluster known as 47 Tucanae.

DOUBLE AND MULTIPLE STARS

β (beta) Tucanae is a multiple star. Binoculars show it as a wide double of 4th and 5th mags., while in small telescopes the brighter component is itself seen to be a pair of near-identical blue-white stars.

δ (delta) Tucanae is a 5th-mag. star with a 9th-mag. companion for small telescopes.

κ (kappa) Tucanae is a double star for small telescopes consisting of components of 5th and 8th mags.

GLOBULAR CLUSTERS

47 Tucanae (NGC 104) is regarded as the second-finest globular cluster in the sky, beaten only by ω (omega) Centauri. It is visible to the naked eye as a hazy 4th-mag. star. Binoculars clearly show the increase in brightness towards its star-packed core, while telescopes of 100 mm aperture begin to resolve the brightest of its 200,000 or more individual stars. The cluster covers nearly as much sky as the full Moon and is a showpiece object for all sizes of instrument. It is a relatively close globular, 17,000 l.y. away.

▶

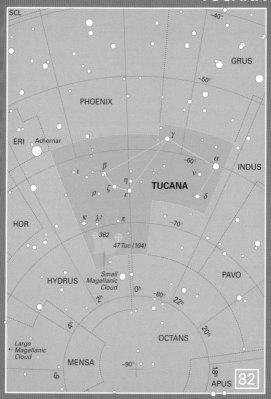

TUCANA

SCL

GRUS

−40°

−50°

PHOENIX

ERI

Achernar

γ

α

−60°

INDUS

ν

β

η

TUCANA

δ

ζ

ρ

ε

ι

κ

λ²

π

−70°

HOR

⊕ 362

47 Tuc (104)

PAVO

Small Magellanic Cloud

HYDRUS

2ʰ

0ʰ

−80°

22ʰ

20ʰ

Large ← Magellanic Cloud

4ʰ

OCTANS

18ʰ

MENSA

−90°

APUS

6ʰ

82

NGC 362 is a 7th-mag. globular cluster visible in binoculars at the edge of the Small Magellanic Cloud. It is
not in fact part of it, but is a foreground object lying
29,000 l.y. away in our own Galaxy.

GALAXY

The Small Magellanic Cloud is the smaller of the two
companion galaxies of our Milky Way. It is also the
more distant, lying about 200,000 l.y. away. It is visible
to the naked eye as a misty patch over 3° long, shaped
somewhat like a tadpole. Binoculars and small telescopes resolve individual stars, clusters and glowing
nebulae in the Small Magellanic Cloud.

*Tucana, the Toucan, as depicted by Johann Bode on his star
atlas of 1801. Royal Greenwich Observatory*

47 Tucanae, a famous globular cluster, has a diameter of about 150 light years and contains hundreds of thousands of stars.
AURA/NOAO/NSF

URSA MAJOR The Great Bear

The third-largest constellation, lying in the northern hemisphere of the sky. It contains a group of seven stars that make up the familiar figure of the Plough or Big Dipper. The stars β (beta) and α (alpha) Ursae Majoris, both 2nd mag., point to the north pole star, Polaris.

DOUBLE AND MULTIPLE STARS
ζ (zeta) Ursae Majoris (Mizar), mag. 2.2, is one of the most famous multiple stars in the sky. Keen eyesight or simple binoculars show a 4th-mag. companion, Alcor (80 Ursae Majoris). In small telescopes another 4th-mag. star is visible closer to Mizar. Each of these stars is also a spectroscopic binary.

ξ (xi) Ursae Majoris is a pair of 4th- and 5th-mag. yellow stars that orbit each other every 60 years. They are currently divisible by apertures of 75 mm and high magnification.

GALAXIES
M81 (NGC 3031) is a 7th-mag. spiral galaxy that appears in small telescopes as an elliptical patch with a noticeably brighter core. Photographs show it to be one of the most beautiful spirals in the sky. Under good conditions both M81 and its partner M82 are visible in binoculars. ▶

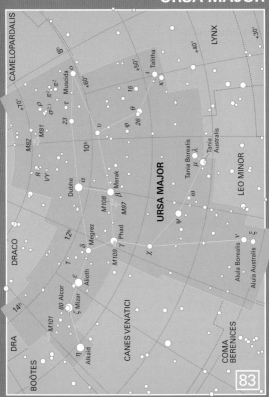

M82 (NGC 3034) is an 8th-mag. galaxy near M81 (see previous page) presented edge-on to us so that it appears as an elongated smudge in small telescopes.

M101 (NGC 5457) is an 8th-mag. spiral galaxy, face-on to us. Photographs show it to be an impressive galaxy with far-flung spiral arms, but only the brightest central part is visible in small telescopes.

URSA MINOR The Little Bear

This constellation contains the north celestial pole.

NORTH POLE STAR
α (alpha) Ursae Minoris (Polaris) is a cream-coloured supergiant of mag. 2.0 that lies within 1° of the north celestial pole. Precession will take it closest to the true pole in about AD 2100. Polaris is a double star, with an 8th-mag. companion visible in small telescopes. It is also a Cepheid variable, but its variations faded out during the 20th century and now amount to only a few hundredths of a magnitude. It is not known whether its variations will start to increase again.

DOUBLE STAR
γ (gamma) Ursae Minoris is a 3rd-mag. white giant with an unrelated 5th-mag. orange giant, 11 Ursae Minoris, nearby, visible to the naked eye or in binoculars.

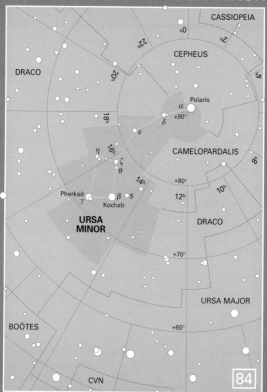

CASSIOPEIA

CEPHEUS

DRACO

Polaris

α

δ +90°

ε

CAMELOPARDALIS

η

ζ

θ

+80°

Pherkad

γ

β 5

Kochab

URSA MINOR

DRACO

+70°

URSA MAJOR

+60°

BOÖTES

CVN

84

VELA The Sails

Formerly part of the ancient constellation of Argo
Navis, the ship of the Argonauts; the other parts are
Carina, Puppis and Pyxis. Vela lies in a rich part of the
Milky Way.

DOUBLE AND MULTIPLE STARS

γ (gamma) Velorum is a binocular duo of 2nd- and 4th-
mag. stars, the brighter of which is the brightest star of
the Wolf–Rayet type (see p. 170). γ Velorum also has
wider companions of 8th and 9th mags., making this an
interesting multiple star for small apertures.

δ (delta) Velorum is a tight double star for apertures of
at least 100 mm, consisting of components of 2nd and
5th mags.

OPEN CLUSTERS

IC 2391 is a bright knot of stars, visible to the naked
eye but best seen in binoculars, scattered around the
4th-mag. star o (omicron) Velorum.

NGC 2547 is a binocular cluster of about 80 stars of
7th mag. and fainter.

PLANETARY NEBULA

NGC 3132 is a large planetary nebula somewhat
smaller than the more famous Ring Nebula in Lyra but
over twice as bright. In telescopes it appears as a hazy,
elliptical disk of similar apparent size to Jupiter.

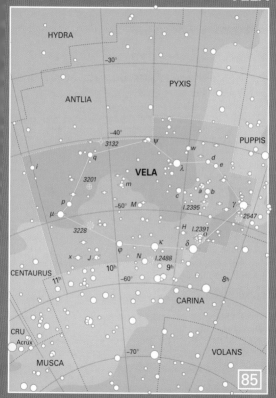

VELA

HYDRA

ANTLIA

PYXIS

PUPPIS

−30°

−40°

3132

ψ

w

d

e

VELA

q

i

3201

m

λ

c

a *b*

p

M

I.2395

γ

−50°

2547

μ

3228

H

I.2391

o

δ

x

J

φ

N

κ

I.2488

9h

CENTAURUS

11h

10h

−60°

8h

CARINA

CRU

Acrux

−70°

VOLANS

MUSCA

85

VIRGO The Virgin

A constellation of the zodiac, in which the Sun lies from mid September to early November. The Sun is in Virgo at the time of the autumnal equinox, i.e. the time when the Sun moves into the southern celestial hemisphere; this happens on September 22 or 23 each year. Virgo is the second-largest constellation in the sky. In various legends she represents the goddess of justice or the goddess of the harvest.

BRIGHT STAR
α (alpha) Virginis (Spica, 'ear of wheat') is a blue-white star of mag. 1.0, lying 260 l.y. away.

DOUBLE STARS
γ (gamma) Virginis consists of a pair of 4th-mag. yellow-white stars that orbit each other every 169 years. As seen from Earth they are closest together around AD 2005 when they will be indivisible in all but the largest amateur telescopes; thereafter they will widen out again, becoming divisible by 75 mm in 2010.

θ (theta) Virginis is a pair of 4th- and 8th-mag. stars for small telescopes.

τ (tau) Virginis is a 4th-mag. star with a wide, unrelated 10th-mag. companion for small telescopes.

▶

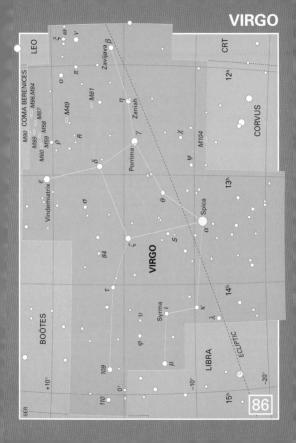

VIRGO

LEO

COMA BERENICES

CRT

CORVUS

BOÖTES

LIBRA

VIRGO

M90 M89 M86,M84 M87 M60 M59 M58 M49 M61 M104

Vindemiatrix
Zavijava
Zaniah
Porrima
Spica
Syrma

ECLIPTIC

SER

86

GALAXIES

Virgo contains a major cluster of galaxies lying about 55 million l.y. away. The Virgo Cluster of galaxies spills over the border into neighbouring Coma Berenices, so sometimes it is known as the Virgo–Coma Cluster. The chart opposite shows the brightest members of the cluster, and the most interesting members are described below. These can be located as faintly glowing patches of light in small telescopes.

M49 (NGC 4472) is an 8th-mag. elliptical galaxy.

M58 (NGC 4579) is a 10th-mag. spiral galaxy.

M60 (NGC 4649) is a 9th-mag. elliptical galaxy.

M84 (NGC 4374) and M86 (NGC 4406) are a pair of 9th-mag. elliptical galaxies visible in the same field.

M87 (NGC 4486) is a 9th-mag. giant elliptical galaxy, also known as the radio source Virgo A. A giant black hole is thought to lie at the heart of this powerful galaxy.

M104 (NGC 4594), the Sombrero Galaxy, is not a member of the Virgo Cluster but lies about 20 million l.y. closer. It is an 8th-mag. spiral seen nearly edge-on, appearing somewhat like a sombrero hat on photographs – hence its popular name.

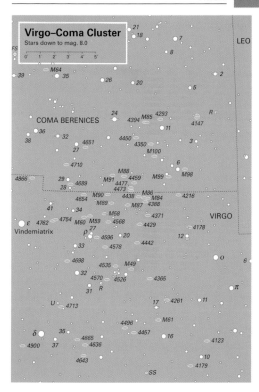

Virgo–Coma Cluster

Stars down to mag. 8.0

0'   1'   2'   3'   4'   5'

VOLANS The Flying Fish

A small and faint constellation in the south polar region of the sky, originally known as Piscis Volans. Its brightest stars are of only 4th mag.

DOUBLE STARS

γ (gamma) Volantis is an attractive pair of golden-yellow and cream stars of 4th and 6th mags. for small telescopes.

ε (epsilon) Volantis is a 4th-mag. blue-white star with a close 7th-mag. companion visible in small telescopes.

Volans, shown under its original title of Piscis Volans, on Johann Bode's 1801 star atlas. Royal Greenwich Observatory

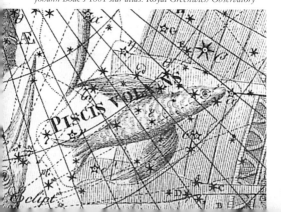

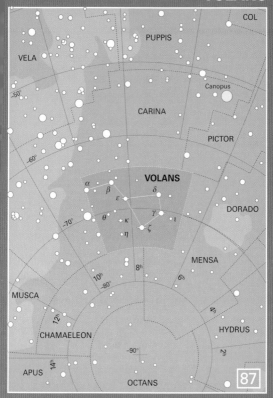

VULPECULA The Fox

A faint but far from uninteresting constellation in the
northern hemisphere of the sky. It was originally known
as Vulpecula cum Anser, the Fox and Goose. In
Vulpecula the first of the flashing radio stars known as
pulsars was discovered by radio astronomers at
Cambridge, England, in 1967.

DOUBLE STAR
α (alpha) Vulpeculae is a 4th-mag. red giant with an
unrelated 6th-mag. companion, 8 Vulpeculae, visible in
binoculars.

OPEN CLUSTER
The Coathanger is a binocular group of stars with an
amusing shape, situated along the border with Sagitta.
Six stars of 6th and 7th mag. form a straight line, and
another four stars make a hook that completes the coathanger shape (see inset on map). The group is also
known as Collinder 399 or Brocchi's Cluster.

PLANETARY NEBULA
M27 (NGC 6853), the Dumbbell Nebula, is a large and
bright planetary nebula, reputedly the most conspicu-
ous of its kind. It is visible in binoculars and small telescopes, appearing as an extended misty green glow
about one-quarter the width of the full Moon.
Photographs show that it has a double-lobed shape,
from which it takes its popular name.

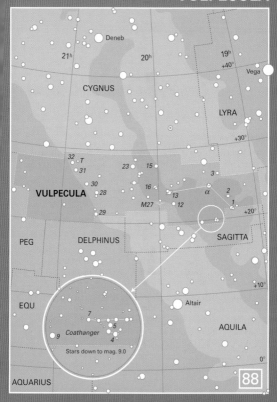

VULPECULA

Deneb

21ʰ

20ʰ

19ʰ

+40°

Vega

CYGNUS

LYRA

+30°

32 ○T

23   15

3

31

VULPECULA

30

16

α

2

28

13

1

29

M27   12

+20°

PEG

DELPHINUS

SAGITTA

+10°

EQU

7

Altair

Coathanger

5

AQUILA

9

3

Stars down to mag. 9.0

0°

AQUARIUS

88

Index

Page numbers in bold denote photographs or diagrams.